The Daily Telegraph B

The Daily Telegraph Book of the Weather

Past and Future Climate Changes Explained

Philip Eden

continuum
LONDON • NEW YORK

Continuum
The Tower Building
11 York Road
London SE1 7NX

370 Lexington Avenue
New York
NY 10017-6503

www.continuumbooks.com

First published 2003

British Library Cataloguing-in-Publication Data
A catalogue record for this book is available from the British Library.

ISBN 0–8264–6197–2 (hardback)
ISBN 0–8264–6698–2 (paperback)

Typeset by YHT Ltd, London
Printed and bound in Great Britain by Biddles Ltd, *www.biddles.co.uk*

Contents

Foreword

Changing climate or no, there is one sentiment which pervades our feelings about the weather during our lifetimes: it isn't what it used to be. How many times have you heard remarks like 'Summers aren't what they were when we were children', 'We never seem to get old-fashioned winters any more', or 'The seasons are all over the place these days'.

Take this comment about a wet summer by a well-known travel writer: 'As for the weather, that will never clear up: eternal rain, clouds and chill. Surely summers were different formerly; or is youth the season of sunshine? For then I thought the summers bright and warm; but now in my age they appear cold and cheerless.' On many an editorial page I read similar comments during the British summers of 2002, 2000 and 1998: but these words were written in August 1787 by John Byng (later Viscount Torrington). Ten years earlier Horace Walpole despaired of another abysmal summer: 'We are in truth but Greenlanders and ought to conform to our climate; we should lay in store provisions and candles for ten months of the year and shut out our twilight ...' Similar sentiments were expressed by Dean Swift early in the eighteenth century, and by the diarists John Evelyn and Samuel Pepys in the 1660s.

So history shows us that the British attitude to our climate has not changed; it is one of complaint occasionally laced with derision and disbelief. There seems to be an inability or an unwillingness to understand that unusual seasons are part and parcel of this country's – any country's – climate. We should also be aware that the human memory is a notoriously poor means of recalling weather events. Extremes tend to stick in the memory, so we remember floods and droughts, snowstorms and damaging gales, heatwaves and intense cold. Hot summers like 1959, 1976 and 1995 stick in the memory, as do severe winters such as 1947 and 1963, but we do not retain specific memories of warm winters or poor summers. Americans and Canadians have been known to complain about their climate too, although they rarely match the professional whingers on the Old World side of the Atlantic Ocean. New Zealanders, Australians and South Africans regularly moan about the rain (though rarely within earshot of an Englishman, to whom their sunny southern hemisphere climates are advertised as unimpeachable).

Statistics cannot give us a complete picture of what a particular summer or winter was like, but they do provide an objective historical perspective which allows us to compare one season with another. Byng's miserable August was actually the warmest for four years and the driest for seven, but Walpole's complaint about June 1777 had more justification: it was the coldest and wettest since 1770. The summer of 1665 drew some disapproving comments from Pepys but the records, albeit very limited, indicate that it was no worse than the previous four or five.

None of these British summers was remotely as bad as that of 1954 – the worst in the last half century. Such was the outpouring of speculation in the newspapers that year that the editor of *Weather*, the Royal Meteorological Society's house magazine, was moved to write: '... there have been such summers before. Situated as we are at the downwind end of the Atlantic, the summer's weather is not so unlikely that atomic explosions, flying saucers, or condensation trails from jet aircraft need be invoked in explanation.'

The purpose of this volume is to provide a comprehensive historical context within which present climatic events and trends can be viewed, and to offer a sound, sober, scientific background against which sensationalized reporting and politicized debating about present and future climate change can be considered. Much is known about the way our planet's weather machine works, but like other scientific disciplines, the more we know the more we realize we don't know. There are many in our society with a little learning who are inclined to sound off mightily, as if they could ever hope to emulate the wide intellectual grasp of our best scientists. All too often these haranguers are the ones listened to because of their eloquence rather than their knowledge. They are more attractive material for the media outlets of the twenty-first century than the more modest and less articulate men and women who populate our research institutes.

Speculation about the future can only be that, but sensible people would prefer informed speculation to the other kind. Into that category fall the chapters which look at weather and climate during the twenty-first century and beyond. It would be a mistake to regard these outlines as predictions, but they are not entirely fantasy either. Rather, they should be regarded as painting a picture, in an impressionist style perhaps, of the sorts of changes and events which are possible.

Philip Eden
July 2002

SECTION ONE

Weather and climate in history

Chapter 1: Weather and climate, what's the difference?

From the earliest days of human history climate has controlled where human beings have settled, while weather events have helped deliver the success or failure of those communities.

The difference between weather and climate is important, but it is often poorly understood. Weather is the state of the atmosphere in a particular place at a particular time, especially in respect of its influence on human beings; it includes the grey skies and drizzle we have today, and the snow and high winds we had yesterday. Climate has often, quite wrongly, been defined as 'average weather'. Rather, it is the synthesis of weather over a long period, therefore including extremes as well as averages. It is the fact that the mean July rainfall in London is 50 millimetres (2 inches), but also that once every ten years, on average, it exceeds 100 millimetres (4 inches), while once every century there is no rain at all during the month. Climate has a character which can be recognized, with variations, from year to year; weather just happens from day to day. You could liken climate to the household budget, with its pattern of income and expenditure throughout the year, while weather is equivalent to the loose change in your pocket or purse which you might use to pay for routine items like a mid-morning coffee, and also for unexpected things like a gift for a friend.

It is evident that during a long period of static climate a community is more likely to be successful, whereas changes in the climate have prompted famine, mass migration, and the collapse of civilizations. Extreme weather has influenced history at a local level, causing a poor harvest, for instance, or a serious flood in a regional capital, or high winds sufficient to demolish the village church. But exceptional weather events have also affected history on a grand scale: the delayed Roman invasion of Britain due to storms in the English Channel, for example, Hitler's inability to complete his invasion of Russia courtesy of two of the severest winters of the twentieth century, and the catastrophic failure of the Americans to rescue their embassy hostages from Iran in April 1981 when dust-storms intervened.

It was believed until the beginning of the twentieth century that the world's climate had not changed during historical times, and probably not for very much longer than that. True, Victorian scientists were aware that climatic convulsions had occurred once upon a time leading to the ice ages, but these were regarded as being in the dim and distant past. Since way before the Romans, it was believed, nothing had changed, and, with an unspoken leap of faith, this established climate was expected to continue indefinitely into the future. It was at this time, near the end of the nineteenth century, that climatologists decided upon 35-year (soon adjusted to 30-year) standard reference periods to describe the climate of a particular town or city. Ideally, the period had to be short enough to guarantee a reasonably large selection of sites with an unbroken record, but long enough so that one or two extreme events did not materially affect the overall statistics. For example, a single rainstorm dumped 213 millimetres (8.4 inches) of rain on Boston, Massachussetts, on an August day in 1955; with this storm the mean monthly rainfall for August in the city for 1941–1970 was 91 millimetres (3.58 inches), but this drops to 84 millimetres (3.3 inches) if that one fall is excluded. This is a significant difference, so it can be seen that the choice of 30 years was a compromise, losing some accuracy in the statistics for the sake of getting a good geographical coverage of recording sites with sufficiently long histories. Nevertheless, these climatologists understood that, using a variety of statistical tools, the whole climate of an individual town could be derived from those 30 years of data – from calculating the highest temperature likely to be recorded in 100 years to assessing how often, on average, a daily rainfall of, say, 200 millimetres (8 inches) was likely to recur.

These statistics are important because they provide information about how the climate impinges on human life: on agricultural production, on water supply, on electricity generation, on flood protection, on the transport infrastructure, and on civil engineering projects. In technologically sophisticated countries such as those of North America, Europe and Australasia, detailed understanding of these averages and extremes saves time and money; in communities surviving on the margins, such knowledge can mean the difference between success and failure, between life and death. If such a society knows that a harvest will fail, perhaps due to excessive rain and inadequate warmth, in one year in twenty, on average, food stores can be kept to cope with such an episode. If the population expands into a more marginal region where harvest failure occurs in one year in five, on average, much more effort will have to be put into maintaining emergency stores to avoid periodic famine. A new strain of a particular crop might be introduced which has a higher yield, but which is more sensitive to high rainfall or low temperatures: is the higher yield worth the increased likelihood of an unsatisfactory harvest?

Belief in a static climate was steadily eroded in the scientific community as researchers discovered many ways to piece together evidence of past climatic change, gradually refining their techniques over the decades. But in the wider world it held sway until at least the 1960s. This is illustrated by UNESCO's *History of Mankind*, published in 1963, which claims: '... by 5000 BC ... the climate, the distribution of vegetation, and all the related factors had settled to approximately their present condition.' Thirty years later, however, UNESCO was at the forefront of publicizing the dangers of climate change, whether natural or induced by human beings.

Climatologists, geologists and glaciologists, meanwhile, were gradually putting together a chronology of climatic fluctuations during the last million years. German and Scandinavian scientists were at the forefront of research concerning the glacial periods, that is before about 9000 BC, while the British climate experts, Charles Brooks and Gordon Manley, studied post-glacial changes. A reasonably complete but fairly crude reconstruction of glacial and post-glacial climate in Europe was available to readers of the scientific journals by 1935, and during subsequent decades this has been progressively refined, and extended to other parts of the world.

Chapter 2: Building a picture of the past

The question that immediately springs to mind when the boffins talk about severe winters in the 1400s, let alone wet periods in the stone age, is: how can they possibly know? After all, barometers and thermometers and rain-gauges were only invented in the seventeenth century, and systematic recording of the vagaries of the weather only began in the middle of the eighteenth century, and then only at a few isolated locations in Europe. It was not until the early 1900s that proper standards for recording temperature, rainfall, sunshine and wind were applied on an international basis, and even then some countries decided to go their own merry way.

Before the instrumental period a completely different approach is needed, and this involves bringing together a wide variety of what are called 'proxy' data. We have already noted that climate affects us in many ways in our everyday lives – it always has done – and it has also always been intimately linked with the character and behaviour of the natural fauna and flora in all parts of the world. What is required, then, is to turn this around and translate the behaviour of human beings and the distribution of plants and animals back into notional climate data. Climatologists have therefore collaborated with historians and economists, geologists and geographers, botanists and zoologists, oceanographers and archaeologists, vulcanologists and chemists, to set out upon this massive task of reconstructing past climates. In recent decades a new breed of researcher, the palaeoclimatologist, has emerged; he or she is trained in several of these disciplines and can therefore follow different lines of research more easily.

Before the beginning of instrumental records in the 1700s, the climate researcher should first of all collaborate with historians. Ships' logs include invaluable records of weather on regular shipping routes back to the middle of the seventeenth century, with intermittent and localized data from much earlier than that. Similarly, the habit of keeping weather diaries was quite common amongst middle-class folk – clergymen, doctors, and politicians, for example – during the sixteenth and seventeenth centuries, not just in Europe but in the American colonies too, and one outstanding register kept by an English priest dates back to 1337–1344. Ordinary journals, including those of famous diarists such as Samuel Pepys, also contain a wealth of descriptive information, especially of extreme weather events or exceptional seasons, while economic historians are able to extract a lot of climatic information across much of Europe from farm and estate records, dates and yields of harvests and vintages, the price of foodstuffs and other essentials, dates of freezing and re-opening of rivers and harbours, military records, and various other historical documents. Taken together, this amazing variety of information has allowed climatologists to complete a chronicle of important climatic events across much of Europe back to 1066. Sporadic information has been gleaned for earlier periods too, particularly during Roman times. These many types of information are, of course, also available during more recent times, when they can be calibrated by reference to concurrent instrumental records.

Dendrochronology is the name given to the dating of past events by reference to tree rings, but botanists show us that the rings themselves contain much information about tree growth each year, and this in turn reveals significant climatic information including the length of the growing season, and rainfall during spring and summer. So far, we have barely scratched the surface of the potential of this source of data, but already overlapping records – recently cut trees, timbers in old buildings, tree trunks recovered from bogs and lakes where the water prevents decomposition – extend our knowledge back to 2000 BC in parts of Europe, and to 6000 BC in the south-western USA. It may well be possible ultimately to produce a complete record back to the end of the last glaciation for certain locations in Europe, the USA, Chile, New Zealand, and south-eastern Australia.

Similar techniques are used to build a chronology of mud deposits (called 'varves') in lakes, river estuaries, and on the continental shelf parts of the oceans. These reveal information about river flow which is in turn dependent on rainfall, and in Sweden, Japan and the USA varve records beginning almost 10,000 years ago have already been analysed. Again, much work remains to be done. The ice-sheets of Antarctica and Greenland also lend themselves to this sort of investigation, and researchers drilling through the Greenland ice-cap in the early 1990s succeeded in retrieving ice-cores all the way to the bedrock; these cores reveal annual layers whose thickness varies according to the quantity of snowfall covering 250,000 years. Chemical analysis of these layers to determine the concentration of the stable isotope, oxygen-18, provides additional information about annual temperature levels.

Zoologists and marine biologists help climatologists to study insect populations, pollen distribution, sea corals, and the remains of tiny marine creatures, all of which are climate-dependent, and all of which have been preserved in sedimentary deposits. These provide, generally speaking, a fairly crude method of studying broad changes in the climate, but they do allow us to fix major climatic shifts in the last 500,000 years to an accuracy of 100 years or better, and back a million or more years to an accuracy of 1000 years.

Dating events by reference to fossilized remnants of flora and fauna is a complex subject involving the decay of radioactive isotopes of various elements which were once part of those plants, animals and insects. The best known of these is the one called carbon-14 which is created by the impact of cosmic rays from outer space on nitrogen atoms. Carbon-14 has an atomic weight of 14, the same as nitrogen, whereas ordinary carbon has an atomic weight of 12. This radioactive carbon enters the Earth's carbon cycle and is therefore found in all living things, and the proportion of carbon-14 is maintained as long as they are alive because they are constantly replenishing it through nutrition. Once dead the replenishment ceases, and the carbon-14 decays back into nitrogen at a known rate, so the ratio of carbon-14 to carbon-12 in the remains of organic material should tell us how many years ago death occurred. The rate of decay of radioactive isotopes is nearly, but not quite, constant; this method of dating has to be fine-tuned by reference to known dates such as those derived from tree rings and Arctic ice-layers.

Astronomers, too, have made an important contribution towards piecing together our picture of the past, and they have also helped us to understand how the climate machine works, because astronomical factors are actually responsible for some of the very long-term fluctuations in world climate. The cyclical nature of several of these astronomical factors make them predictable, so they provide a regular underlying groundswell upon which a multitude of shorter-term fluctuations are superimposed – a sort of sostenuto, played with an almost unbearably slow rhythm, beneath the complex music of climatic variability.

There are three primary astronomical cycles whose signatures can be detected in climatic records: orbital eccentricity, axial obliquity, and precession of the equinoxes. Orbital eccentricity describes the change in the shape of the Earth's orbit around the sun from almost circular to more elliptical and back again, a cycle which takes between 90,000 and 100,000 years to complete. Axial obliquity describes a rhythmical change in the tilt of the Earth on its axis, and this completes a cycle in about 41,000 years. The precession of the equinoxes concerns the gradual change in the time of the year at which the Earth reaches its closest point to the sun; this occurs in January at the moment but is drifting backwards at a rate of approximately one year every 21,000 years (in practice it varies between 19,000 and 23,000 years due to interference by the orbital eccentricity cycle). These three phenomena change the amount and distribution of heat energy from the Sun received at the Earth's surface. Climatologists estimate that over 80 per cent of the alternations between glacial and inter-glacial periods during the last million years can be explained in terms of these long-term astronomical cycles. It is evident that they will also be useful in predicting a substantial part of future climate change too, as will be seen in a later chapter. We should note here that there are further astronomical influences on climate, including the regular and entirely predictable diurnal (day-night) and annual (summer-winter) cycles, semi-regular fluctuations in solar output on a scale of hundreds or thousands of years, the impermanent 11.3-year sunspot cycle, and effectively unpredictable events such as asteroid and comet impacts.

Chapter 3: The end of the last glaciation

The change in world climate at the end of the last glacial period is peculiarly important to us at the beginning of the twenty-first century because the accelerated warming trend apparent during the last two decades has been more rapid, according to a majority of palaeoclimatologists, than any in the last 11,000 years.

To avoid confusion, some definitions are appropriate here. The last time ice extended across large parts of Europe and North America is colloquially known as 'the last ice age': but to the specialist we are still in an ice age. Here is what the *Meteorological Glossary* of the UK Met Office says:

> Ice-age: A geological period during which great glaciers and ice sheets extend from the polar regions to as far equatorwards as about latitude 50 degrees. Individual ice-ages wax (glacial phases) and wane (interglacial phases) in strength. Present-day climate is considered to be that appropriate to an interglacial phase of the Quaternary Ice-Age which began in the northern hemisphere about two to three million years ago; since then, the hemisphere has suffered predominantly glacial conditions for about 90 per cent of the time. Glacial phases lasting about 100,000 years are separated by much warmer interglacial phases lasting about 10,000 to 12,000 years.

The ice reached its greatest extent in both Europe and North America during the last-but-one glacial phase roughly 150,000 years ago, but the most recent glaciation was almost as severe. Known as the *Wisconsin* in the USA and Canada, the *Weichsel* in western and northern Europe, the *Würm* in the Alpine region, and the *Devensian* in the UK, it was at its most intense from 24,000 to 14,000 years ago with the greatest extent of ice – the glacial maximum – some 21,000 years ago. There is no international consensus for naming the various glacial and interglacial episodes, but most modern texts combine the American and European names, making the last glacial phase the *Wisconsin-Weichsel.*

In North America the furthest limit of ice reached a line running roughly from where Seattle now is along the 47th parallel to North Dakota, then south-east to Iowa, Illinois and Indiana, and across Ohio and Pennsylvania to Long Island. Canada and Alaska lay entirely under the ice, as did Hudson Bay, but Davis Strait was open water as far north as the latitude 77 degrees. South of the ice-sheet there was a narrow zone of permafrost, generally less than 500 kilometres (300 miles) wide, there was no significant outlying ice-cap over the southern Rockies, and much of the present day USA away from Florida, the Gulf coast and lowland California was tundra.

In Europe the ice-front extended from the Arctic Circle in Russia (there was not sufficient snowfall over Russia for the ice to build up further south) to northern Poland and northern Germany, thence northward into western Denmark, and across the central North Sea to Britain. It followed a contorted course across the British Isles, paralleling the (present-day) east coast from Cromer north to Teesside, then west to the Pennine foothills, south to the Peak District, west again to the Welsh Marches, south again to Glamorgan and across to Pembroke; in Ireland the ice reached a line from Dublin to the Shannon estuary. There was a broad zone of permafrost over 1600 kilometres (1000 miles) wide stretching southward to roughly latitude 45 degrees across Europe, and a substantial separate ice-cap covering the entire Alpine massif. Iceland, too, lay beneath its own ice-cap.

In the southern hemisphere there were permanent ice-caps over the Andes in southern parts of Chile and Argentina, over the interior of Tasmania, and over the Australian Alps (probably including where Canberra now stands), and glaciers existed over the higher mountain slopes in Papua New Guinea. The massive amount of water locked up in the ice-sheets meant that sea-level was very much lower than it is now – at glacial maximum approximately 110 metres (360 feet) lower.

Temperature estimates are rather speculative, but in southern England the average January figure was probably close to −30°C (−22°F) compared with 4°C (39°F) now, and the average July figure about 7°C (45°F) compared with the present day 17°C (63°F). Similar temperatures probably obtained in other regions just beyond the ice-edge such as central Europe, the eastern seaboard of the USA between New York and Washington, and, on the west coast, the lowlands between Seattle and Portland.

The end of the last glacial period occurred between 14,000 and 10,500 years before the present day, but it was not a gradual change; rather it occurred in fits and starts, and there was a marked reversal between 12,000 and 11,000 years ago which has been identified in most parts of the world, but which was especially sudden and dramatic in the Atlantic-Europe region. This relatively brief cold episode is associated with a re-advance of glaciers in Scotland and Scandinavia and is generally known as the *Younger Dryas*, after a group of Arctic and Alpine plants belonging to the genus *Dryas*. In the UK it is sometimes called the *Loch Lomond stadial*. For a long time the Younger Dryas presented a conundrum to the researchers: the astronomical cycles had come together to deliver more energy from the sun and thus initiate the end of the glacial period only a few thousand years before, and solar radiation levels were actually higher than they are today – in other words we were getting more heat from the sun then than we are now. There was, therefore, no obvious explanation either for the sudden return of Arctic conditions, or for their equally rapid retreat less than a thousand years later. It is possible that both the deterioration and the subsequent amelioration were each squeezed into a few decades, which suggests a response to a specific event or series of events on Earth rather than to a long-term cycle.

The most favoured theory now propounded by the experts involves the gradual decay of the Laurentide ice-sheet over Canada. The geography of the Canadian shield, bordered to the west, south and east by high ground, and blocked by ice to the north, meant that melt-water had no escape route. Vast lakes formed, of which the present-day Great Lakes, Lake Winnipeg, and so on are but small remnants. Ultimately, the ice which dammed the 160-kilometre (100-mile) wide passage between Baffin Island and the mountains of northern Labrador was breached, cold fresh water and icebergs poured forth into Baffin Bay and thence into the northern Atlantic Ocean. We know that about this time worldwide sea-level rose by some 40 metres (130 feet) in very short order, and we also know that surface waters of the Atlantic were colder at this time than they were at the peak of the preceding glacial phase. The impact on the climate of the region would have been almost immediate, and this would also explain why the sudden cooling appeared to affect western Europe more than any other part of the world. The changes are illustrated by examining again estimated temperatures in southern England. At 13,500 years before present, during the initial warming, Januarys averaged between 0 and 4°C (32–39°F) and Julys between 12 and 16°C (54–61°F); at 11,500 years before present, in the midst of the Younger Dryas, Januarys were −16 to −20°C (+4 to −4°F) and Julys 8 to 12°C (46–54°F); 10,000 years ago, after the Younger Dryas had ended, Januarys were again around 0 to 4°C (32–39°F), while Julys had rebounded to 14 to 18°C (57–64°F). Climatic changes of such magnitude within a period of, say, 50 years, would be catastrophic were they to occur today.

Once the supply of ice and meltwater flowing out of Hudson Bay from the Laurentide ice-sheet had been exhausted, the cold surface waters in the Atlantic dissipated, and the climate reverted to a much warmer character. At this point, roughly 10,500 years ago, or 8500 BC, the last glacial phase ended and the present interglacial phase began. However, the events of the Younger Dryas may have implications for future climate change as we shall see in a later chapter.

Chapter 4: Post-glacial climates, 8500 BC to AD

Even though the climate of most mid-latitude regions in North America, Europe and Asia had recovered almost to present-day levels by 8500 BC, ice was still melting in Scandinavia and Canada, and as a consequence sea-levels continued to rise until about 2500 BC, the rises coming in spasms, and the geography of the world periodically re-arranging itself. Initially, there were land bridges where none exists today, for instance, between Siberia and Alaska, between the British Isles and the European mainland, across the Bosphorus, connecting several of the Indonesian islands and the Malayan peninsula, and between Australia and New Guinea. These were important for the migration of many species including human beings during the first few thousand years of the present interglacial. However, the principal human migrations from Asia to the Americas and to Australia began very much earlier, probably during somewhat warmer episodes during the early and middle parts of the *Wisconsin-Weichsel* glaciation.

Very nearly all of these land bridges had gone by 7000 BC. A thousand years earlier the whole of what is now the southern North Sea was a broad, ill-drained plain, and the Thames was merely a tributary of the Rhine which disgorged its contents into the sea somewhere to the west of a range of low hills which now form the Dogger Bank. Over 130,000 square kilometres (50,000 square miles) of land vanished in the North Sea area alone during 500 years, and the North Sea and English Channel linked in approximately 7500 BC. It is also likely that many early human communities, typically located in coastal locations, would have perished when rising sea-level combined with severe storms resulting in devastating and extensive floods.

Temperatures continued to climb, albeit at a much slower rate, during the early post-glacial period. Humans followed the retreating ice edge very closely, and there is evidence of groups of hunter-gatherers migrating from Denmark across the North Sea lowlands to north-east England in about 8000 BC, while other peoples hunted reindeer in southern Norway in 7000 BC. Forests gradually invaded the tundra, with a succession of species travelling northwards in waves, both in Europe and in North America. Stunted birch groves appeared first, soon growing more vigorously, then pine spread widely throughout both continents. As the climate became warmer, a variety of deciduous species supplanted much of the pine forest in middle latitudes, with oak, elm and hazel prominent in Europe, and additionally hornbeam and maple in North America. Lime trees were prominent in drier regions, and elder in areas where rainfall was plentiful throughout the year. Pine forests continued to thrive in colder latitudes as indeed they do today – in Canada, in northern Russia, in Scandinavia – and also on the cold mountain slopes of the Alps and western Rockies. Animals, birds and insects found suitable habitats in more and more places, so they, too, became geographically more extensive and also more numerous.

Changing climate and rising sea-level in the Middle East also precipitated mass migrations away from the disappearing coastal lowlands, and these environmental factors may have contributed to the birth of agriculture here, in the form of arable farming and animal husbandry, probably around 8000–9000 BC. Urban societies in the Near East also first appeared around this time.

Post-glacial warming peaked in most parts of the world between 6000 and 3500 BC in a phase which is sometimes described as the 'Climatic Optimum'. In middle latitudes of the northern hemisphere, including central and northern Europe, the USA north of the 40th parallel, and southern Canada, this warmer episode developed more slowly and lasted longer than in other parts of the globe, culminating between 4000 and 1500 BC – such a level of warmth has not been approached since. Climatic characteristics during this period have been reconstructed using primarily tree-pollen data which is abundant in many regions, supplemented by animal, bird and insect fossils, and also by tree-ring analysis. Globally, mean annual temperature was some 1.5° to 2°C (2.7° to 3.6°F) higher than the twentieth-century average; in northern and western Europe winters were 2° to 2.5°C (3.6° to 4.5°F) warmer and summers 1.5° to 2°C (2.7 to 3.6°F) warmer. The present tree-line in the Scottish highlands stands at around 600 metres (2000 feet) but there is plenty of evidence to show that it lay close to 750 metres (2500 feet) throughout that 2500-year long warm period, and as high as 780 metres (2600 feet) at 3000 to 3200 BC. Analysis of similar records in northern Canada shows that the extreme poleward limit of coniferous forest was reached in two distinct phases, one around 4000 BC and the second near 2500 BC.

The broadscale atmospheric circulation was considerably distorted during the Climatic Optimum compared with the present day. The last significant remnants of Scandinavian ice had wasted away by 6000 BC and the Laurentian ice-sheet had almost completely melted a thousand years later, but the Baltic Sea and Hudson Bay areas which had been weighed down by the ice now became extensive arms of the Atlantic Ocean – much larger sea-areas than they are today. While there remained extensive ice over the northlands of Canada and Scandinavia, high pressure systems dominated these areas even in summer (domes of cold air persisted over the ice, cold air is dense, so barometric pressure remains high), diverting the main depression track well to the south across the USA, then north-north-eastwards across the western Atlantic, thence dividing either side of Greenland; meanwhile the semi-permanent Azores 'high' was displaced northeast of its present position and often extended across western and central Europe and the British Isles. In winter, the Atlantic storm track was again displaced well to the north of its present position while high pressure was found in mid-Atlantic and over central Europe more often than now.

The overall effect on the European climate was that it was not only warm, but it was also relatively dry – mean annual rainfall is estimated at 15 to 20 per cent lower than the present-day figure – and also markedly less stormy than it is now. This is at odds with computer model predictions for a twenty-first century warming trend which indicate that higher temperatures in the Atlantic-European sector are likely to be accompanied by higher rainfall and increased storminess. However, it should be pointed out in fairness to our research climatologists that they emphasize our computer models are not yet sufficiently refined to put any reliance on the rainfall trends that they produce. Furthermore we are not, in the twenty-first century, emerging from a glacial period, ocean temperatures are appreciably higher, and so it would be wrong to try to draw parallels. Some of the flora and fauna found in southern Britain, the Low Countries, northern Germany, Denmark and southern Sweden at this time could only be supported by midsummer mean temperatures approaching 20°C (68°F), a level which is now indicative of our very hottest summers – say once every ten or twenty years.

The marked poleward shift of the sub-tropical high pressure belt – the Azores 'high' in the Atlantic sector – during this period had a very important knock-on effect closer to the equator. Today, the equatorial zone has heavy rainfall throughout the year while heavy summer rains affect adjacent latitudinal zones to the north and south. Between 6000 and 3500 BC these seasonal rains affected much broader belts either side of the equatorial zone. We certainly know that this is true of the Sahara, of Arabia and Mesopotamia, and of the Indus Basin in what is now Pakistan, and there is every reason to believe that the same could be said for the south-western USA, and the deserts of Australia and southern Africa.

For almost three thousand years the Sahara Desert was no desert at all but a vast savannah with grasslands, some woodland, rivers and huge lakes. It is estimated that much of the region received between five and ten times the present average annual rainfall. The present Lake Chad is but a tiny remnant of a huge expanse of water which at its peak was larger than the Aral Sea is now. Rock-paintings and early hieroglyphs indicate that mesolithic (Middle Stone Age) hunter-gatherers were numerous throughout the Sahara region, and that they hunted buffalo, rhinoceros, elephant, and giraffe, and some of these paintings also depict boats, fish, hippotamus and crocodile, suggesting that some subsisted in an environment dominated by water. After about 4000BC the character of the pictures and pictograms changed, now showing domesticated cattle. Lands bordering the Nile were fertile and supported a burgeoning population in the early fourth millennium BC, before the founding of the Egyptian dynastic civilisation. Neolithic (New Stone Age) populations were also growing around the Tigris and Euphrates rivers in what is now Iraq, and along the Indus in Pakistan where four to eight times the present-day rainfall was characteristic of the period between 8500 and 2000 BC. Here, too, rhinoceros and water buffalo roamed; dates, cotton and melons were cropped; today it is one of the most inhospitable deserts in the world.

Wholesale changes in climate patterns happened in the middle of the fourth millennium BC, possibly associated with the final collapse of the sub-Arctic high pressure systems over Eurasia and North America; the tundra receded, winter snows diminished, and evergreen forests expanded into regions where no trees have been before or since during the present interglacial phase. Favoured depression tracks over North America and the Atlantic-Europe sector were no longer diverted first far to the south and then far to the north; rather, the depressions travelled more frequently in an easterly direction across Canada and the USA, and took an easterly or north-easterly course across the Atlantic Ocean and Europe, now coming closer to Scotland and often crossing Scandinavia. Western and northern Europe became wetter and windier but remained warm in comparison with today. Forests in Scandinavia retreated slightly in contrast to their Canadian and Siberian counterparts. In Britain the tree-line if anything climbed higher in line with the high temperature levels, but there is much evidence that peat bogs became more extensive in Scotland, northern England and Wales, implying higher rainfall.

The sub-tropical high pressure belt slipped southwards into north Africa, subduing the Saharan summer rains and triggering desertification. Some historians have suggested that the growing aridity of the region was caused by over-grazing as neolithic tribesmen amassed large herds of cattle during the four or five centuries preceding the change. Although extensive grazing did not help matters, the palaeoclimatic evidence for a worldwide climate change at this time has now stacked up so high that there is no need to seek any other cause.

Paradoxically, the growing climatic stress was probably responsible, in part at least, for the development at this time of the great early civilizations. As the Sahara and other deserts grew, great migrations of human populations began. Nomadic peoples from the increasingly dry Sahara region travelled to the fertile lowlands along the Nile river and Mediterranean coasts, while similar concentrations of population occurred along the Tigris and Euphrates, in the Indus Valley, and in the major river valleys in China. As Professor Hubert Lamb suggests: 'If pastures and stocks of wild game for hunting were failing, the advantages of cultivation in more or less reliably irrigated valleys would be more obvious to those faced with abandoning an age-old way of life.' Lamb also suggests that the stories of migration of the forebears of the Israelites in the early books of the Old Testament of the Bible are the nearest thing to a written history of these great population movements that we have.

The development of the great civilizations in these regions, all during the third millennium BC, ensured the development of the agricultural techniques and social practices necessary to support the greatly inflated numbers of people. It is probably true that domesticated animals and cultivated crops of grain, pulses and roots do not provide a greater level of nutrition or variety compared with the diet of hunter-gatherers, but without a doubt agricultural practices did provide many more calories of food for a given area, and for fewer man-hours of work too. Such a change was essential to cope with the growing worldwide human population, which was estimated at five million in 7500 BC and over hundred million by 2500 BC.

In Europe, although the climate remained predominately warm into the second millennium BC, it does seem to have become more variable after 3500BC with higher rainfall. Indeed there appears to have been a brief deterioration lasting 300 or 400 years, with temporarily lower temperatures. Such changes would undoubtedly have had serious repercussions on many human settlements which had enjoyed a generally stable environment for a long period. Rivers changed their courses, animal pasture turned into marsh and bog, and some 'roads' used for travelling and trading became unusable during the winter season. It is again noteworthy that this environmental stress coincided with the rapid spread of cultivation skills across the continent. This was the beginning of the Neolithic period – the Late Stone Age.

Another hiccup in the generally benign conditions in Europe happened around 2200–2000 BC, as the Neolithic period gave way to the Bronze Age and a fresh wave of technological advances brought the first metal tools and jewellery to the growing population. Even at this point, though, the records of tree and grass pollen indicate that northern Europe, southern Canada and the north-eastern USA were all significantly warmer in 2000 BC than in 2000 AD. Striking confirmation of this is found on Dartmoor, in south-west England, where relics of Bronze Age farming have been found 150–200 metres (500–600 feet) higher than the present-day limit of cultivation. The cool period around 3000 BC also shows up clearly in North American records, specifically in tree-ring evidence from California, but the second cool spell around 2000 BC is not detectable.

Around the Arctic fringes of North America, northern Europe and Siberia, the highest post-glacial temperatures were delayed until 2000–1500 BC; but in temperate and sub-tropical latitudes the climate was already changing, as we have seen, and a further marked deterioration took place during the second millennium BC, bringing colder and wetter conditions in middle latitudes and accelerated desertification in the subtropics. After 1900 BC a catastrophic drop in rainfall led to the extinction of the Indus civilizations, and there was episodic drought leading to severe famines in Egypt around 2150, 2000 and 1800 BC.

The Californian tree-ring record shows a marked change to lower temperature and increased rainfall around 1300 BC. In the Scottish highlands the tree-line dropped abruptly from 750 metres to 500 metres (2400 feet to 1600 feet), equivalent to a temperature drop of approximately 1.5°C (2.7°F), around 1200 BC; about the same time in England there was a marked decline in elm forest and a growth of peat-bogs in the lowlands, while Bronze Age settlements in the Pennines and the Welsh hills were abandoned.

Evidence from the southern hemisphere is much more limited, but Australia's climate showed a clear trend to drier and cooler conditions after 2000 BC, and the drop in the mountain tree-line shown in Europe and North America was mirrored in New Zealand and South America. Nor was this planet-wide deterioration a shortlived affair; indeed it is useful to regard this regime as continuing to the present day punctuated by relatively brief warmer episodes. These short warm periods occurred during Roman times, in the early Middle Ages, and during the twentieth century; we will come to them in due course. Indeed, many present day glaciers in the Alps, Scandinavia, the Rockies, the Andes and New Zealand re-appeared between 1500 and 1000 BC, having previously been absent for thousands of years.

We can sketch the probable prevailing weather patterns during the first millennium BC. In temperate latitudes, the storm tracks across the Pacific and Atlantic Oceans were now close to their present-day positions, much further south than they had been during the Climatic Optimum. Heavy rains and damaging gales probably became more frequent in the Pacific Northwest of the USA, British Columbia, the British Isles and north-western Europe, while summer warmth was much less reliable and some summers would have been plagued by a succession of depressions travelling from the oceans between latitudes 50 and 60 degrees north. Lands bordering the North Sea endured a marked increase in the frequency of coastal floods, although sea-level was 1 to 1.5 metres (3 to 5 feet) lower than it had been during the previous millennium.

Around 300 BC the intensity and frequency of these storms reached a peak in Britain, with annual rainfall perhaps 40 to 50 per cent higher than it is now, and most summers were cool and rainy. The Somerset Levels, around the ancient settlement of Glastonbury, had been dry before 1200 BC and were criss-crossed by well-used footpaths, but these flatlands, barely above sea-level, turned to marshland by 1000 BC and thousands of trees were cut and transported to create wooden causeways to maintain trading routes. Even these became impossible to use after 500 BC, being superseded by boats, and the villages themselves at this time were built on piles to keep the houses reasonably dry. Elsewhere in England, hilltop roads such as the Icknield Way became the main communication routes.

Hereabouts the British climate was probably as cool as at any time since the Younger Dryas episode, and temperatures were at least as low as they were during the Little Ice Age of the sixteenth and seventeenth centuries AD. Glacial deposits in Norway, Sweden and the Alpine region confirm this; some glaciers advanced even further at this time than they did in more recent centuries. Evidence of increased windiness in north-west Europe is found in the genesis of extensive sand-dunes and sand-spits, especially between 600 and 100 BC.

Autumn and winter rainfall became more abundant and more reliable during the first millennium BC in the Mediterranean basin as occasional Atlantic storms took a more southerly track, but the deserts of north Africa and southern and central Asia grew more extensive and more arid as the intermittent incursions of equatorial moisture which had characterized the Climatic Optimum finally dried up completely. The central and eastern Mediterranean therefore enjoyed a kinder climate than heretofore, providing a positive background to the rise of the Ancient Greek civilization, and somewhat later to the establishment of Carthage and Rome.

We have already noted the evidence that North America's climate also turned cooler and moister after 1300 BC and this regime lasted for at least a thousand years. The lower temperatures and more reliable rainfall extended to Mexico, and probably to the Central American isthmus as well, and we may note that the first flowering of the civilization here happened between 600 and 100 BC. Subsequent periods of high culture in this region coincided with later cool episodes in middle latitudes of the northern hemisphere. In the far north – Greenland and Arctic Canada – it appears that the cooling process came hard on the heels of the maximum warmth, and it came in two phases with a sharp decline in temperature around 1200 BC and a further decline near 700 BC. Many aboriginal settlements in these Arctic regions were abandoned during these periods, and there is evidence of a general migration of populations southward into northern Ontario, Quebec and Labrador.

The reason for the dramatic downturn around 1300 BC for which evidence has been found in widely spread locations across the world has not yet been pinpointed. Indeed, there may be several reasons. But the coincidence of the change in both hemispheres, in tropical and polar regions, on mountains and sea-shores, suggests that there may have been an outside agency involved. A medium-sized asteroidal or cometary impact has been postulated, or there may have been a sudden spurt of volcanic activity, or possibly the former led to the latter. We know that the volcanic island of Thira (previously known as Santorine) exploded violently in approximately 1450 BC, but the injection of volcanic debris into the atmosphere after a single eruption would cool the climate for only two or three years. It would need something much longer-lasting to initiate the step-change in world climates that we have described here.

Chapter 5: The first millennium AD

The Romans provided us with the first written accounts of weather and climate over a wide area, although occasional narratives – such as those in the Bible – offer some insights into prevailing conditions at individual locations at an even earlier date. The Roman descriptions come from chroniclers such as Livy, Pliny and Ptolemy, travellers such as Pytheas and Tacitus, and from the writings of military men who of course also travelled widely across Europe, north Africa, and the Near East.

Tree-ring and pollen records indicate a slight amelioration in the northern and western European climate, mirrored in North America, after 250 BC. But the emphasis is on the 'slight': winters were still stormy and summers often wet and windy. A succession of violent storms in the North Sea basin is known to have occurred in the two decades before 100 BC, significantly changing the shape of the coastlines of what are now Denmark, north-west Germany and the Netherlands, and large areas were abandoned by the local populations who moved southward and eastward into Germany and Central Europe. A further wave of migrations which brought the Belgic peoples into southern Britain may have been triggered by a similar series of events. And then the Romans' attempted invasion of Britain under Julius Caesar during the consecutive summers of 55 and 54 BC was confounded by long periods of strong westerly and northerly winds which made the English Channel impossible to cross. Such conditions occur in summer nowadays about once every eight to ten years on average.

The slight improvement noted after 250 BC introduced a period of stable climate in north-western Europe which lasted until exactly 400 AD, but southern Europe and northern Africa experienced a gradual drying trend after 100 AD. It is well known that North Africa was the granary of the Roman Empire, but the cultivated fields here, and cities such as Petra in what is now Jordan, were subsequently overwhelmed by the growing deserts. Ptolemy's weather observations in Alexandria early in the second century AD show clearly that the climate of northern Egypt was at that time very different to what it is today. Rain showers occurred throughout the year – rain hardly ever falls now between April and October – while the stiff sea-breezes, which now blow daily during the summer half of the year and provide some relief from the high temperatures, then often failed, resulting in lengthy periods of overpowering heat.

The drier regime in the Mediterranean was at its height between 200 and 550 AD, and this presented problems to the Empire on two fronts. At home, increasing drought resulted in less reliable and poorer harvests, it became more difficult to feed the populace, and there were also recurring bouts of plague and other diseases. The rigid social order and high taxation imposed by Emperor Diocletian in the year 293 may have been a response to these growing environmental stresses. Further afield – and out of sight – the same atmospheric mechanism was reducing rainfall in the already sparsely-watered lands of central Asia, now documented by reference to the changing water-levels of the Caspian and Aral Seas, of Lake Balkhash in modern Kazakhstan, and in the Tarim and Junggar depressions in the present-day Xinjiang (Sinkiang) region of China. Just as the desertification of the Sahara initiated mass movement of peoples almost 4000 years earlier, so this growing aridity in central Asia which turned the steppes to desert triggered a new wave of migrations. High mountains barred the way to the south and east, the inhospitable wastes of Siberia lay to the north, so the only way out was to the west. Invasion of the imperial lands from the east became increasingly frequent during the fourth century, culminating in the sack of Rome in 410.

In Britain and adjacent parts of the continent there is clear evidence of a gradual warming during the Roman occupation, which was associated with a retreat of glaciers in the Alps, a rise in sea-level of one or two metres (three or six feet), and a reduction of storminess, especially in summer. The sea-level change led to the invasion by the sea of the Fen District, the Norfolk Broads, the Rhine estuary and the Flanders coastal plain, probably between the years 270 and 300. The Romans introduced wine-growing to Britain, and we know that vineyards were numerous in south-east England with one or two in the Midlands, East Anglia, and even as far north as Yorkshire. Some have postulated on this evidence that the climate was therefore warmer than it is nowadays when the commercial cultivation of the vine is largely restricted to Kent, Sussex and Hampshire. However, the commercial exigencies and the market characteristics which exist in the early twenty-first century certainly did not apply in Roman times; indeed the present-day climate is quite warm enough to grow the vine in sheltered sites as far north as Cheshire and Durham with the prospect of a reasonable crop five years out of ten. Nevertheless, it is interesting to note that Britain appeared to be self-sufficient for most of the fourth century, with no indication of regular wine imports. Other warmth-loving crops were not successfully introduced, and this confirms our estimation that summers in Britain were no warmer then than they are now. Winters, if anything, were colder with frequent snow and frost, but the Atlantic depression track appears to have been displaced far to the north, and as a consequence winter storminess was suppressed.

A sharp climatic deterioration hit most of Europe early in the fifth century, with sudden cooling in the years 400 and 415. In Britain this coincided roughly with the withdrawal of the Romans and a growing threat of invasion by the Saxons and Angles. There also appears to have been a succession of abnormally cold years around 640 which may have been the result of a series of huge volcanic eruptions or even of a small asteroid impact. From here until the end of the first millennium Britain's climate reverted to its pre-Roman character, with appreciably colder winters and cooler windier summers. European glaciers advanced and the sea-level dropped, but the Fens remained undrained and were occasionally submerged by storm surges during severe North Sea gales. The Somerset Levels, too, returned to marshland for a while, but the monks at Glastonbury set themselves to reclaim the surrounding lands by building drainage ditches and protective dikes. We know of some seriously cold winters in southern Europe and northern Africa during this cool epoch: for instance, the sea froze in the northern Adriatic and around Byzantium (Istanbul) in 764, 860 and 1011.

Research in Canada and Alaska has revealed that glaciers there, too, advanced between the fifth and tenth centuries, and the Californian tree-ring series indicates a gradual climatic deterioration after 400, culminating in a period of exceptional coolness between 800 and 1000 – probably the coldest post-glacial period of all in this particular region.

All these changes are consistent with a narrowing of the equatorial climatic zone, and the coincident shift equatorwards of all the other climatic zones. Thus the sub-tropical high pressure areas retreated from southern Europe bringing seasonal rains back to the Mediterranean, and settling instead over north Africa, the Near East and Arabia where increasing aridity contributed to the decline of city states. So the burgeoning of Islam took place against a background of drought, food shortages, disease, and migrating populations.

The second flowering of civilization in Central America was also associated with these drier centuries: the ruins of the great Mayan cities are now overwhelmed by tropical forest, but they were built mainly on open grassland. The population of the region grew to several million, necessitating an expansion into the coastal lowlands where the retreat of the forests was accelerated by slash and burn. Increasing rainfall during the ninth century encouraged the tropical forests to re-colonize the region at a time when the civilization had begun to decay. It has even been suggested that the final collapse may have been brought about by the catastrophic effects of a great hurricane; this region of Central America is particularly prone to hurricane activity, and the widespread devastation and loss of life caused by Hurricane 'Mitch' in October 1998 may give an inkling of what might have befallen the Maya over a thousand years earlier.

Chapter 6: The last 1000 years in Europe

Weatherwise, it was a millennium just like many others. There were several warm centuries, a cold period lasting three or four centuries, and we had a number of destructive individual events interspersed with much meteorological mundaneness.

Broadly speaking, there were three contrasting episodes: a mediaeval warm period sometimes dubbed the 'Little Optimum' by climate historians; a lengthy cold interlude which culminated between 1550 and 1700 and which is usually referred to as the 'Little Ice Age'; and a relatively rapid warming during the final 150 years of the millennium. Between these dates there was a prolonged but erratic decline between 1300 and 1550, and a slow and fitful improvement between 1700 and 1850.

Good King Wenceslas had looked out for the last time some 66 years before the second millennium began, but his English counterpart, Aethelred II ('the Unready'), would have regularly observed snow lying on the Hampshire fields around his capital city, Winchester, and no doubt it was on occasion fairly deep, quite crisp, and reasonably even. Those fields were more often snow-covered in those early years of the eleventh century because the winter climate was about a degree colder than it is now.

But Britain's climate was already on an upward trend, and the twelfth and thirteenth centuries (along with the twentieth) were the warmest of the whole period. Winters were very variable, usually mild and rainy but with occasional severe ones, springs were often warm and showery and frost-free, summers were consistently good with abundant sunshine along with welcome thundery downpours which kept drought at bay, and autumns were mild but sometimes wet and windy. Crops were harvested at higher altitudes than at any time since the Climatic Optimum – at 400 metres (1300 feet) above sea-level on the fringes of Dartmoor and south-east of Edinburgh in the Lammermuir Hills – and sheep farmers complained that there was too little land left for grazing. Commercially successful vineyards were developed in many parts of southern and eastern England with isolated sites as far north as Lincolnshire and the Vale of York. Several of these continued production for 100 to 200 years. The quality of some English wines was so good that the French tried to negotiate a treaty to have these vineyards closed down. It is no accident that this long period of benign climate, in continental Europe as well as in Britain, coincided with the great period of cathedral building, technological advances, and the Crusades.

Unseen changes were already afoot during the 1200s. The Norse settlers in Greenland had noted a marked advance of the Arctic ice beginning in 1198, and a deterioration continued throughout the 13th, 14th and 15th centuries; regular contact with Europe ceased in 1369 and the largest settlement was finally wiped out around 1500. No major change in temperature was noted in Britain until the early 1300s – indeed the period 1280 to 1311 arguably marked the peak of this medieval warm period – but the first symptoms of change were appearing in the guise of more frequent North Sea gales with consequent flooding of the coastal lowlands both in England and on the continent. The ports of Ravensburgh in the East Riding of Yorkshire and Dunwich in Suffolk were both lost to the sea following the acceleration of coastal erosion.

Then between 1313 and 1321 it suddenly happened. A sequence of cold and wet summers accompanied by heavy spring rains and also prolonged autumn downpours hit the British Isles and much of western Europe. The change must have been as inexplicable as it was devastating. Vineyards failed, grain yields were sharply reduced, and in 1315 the harvest was ruined by incessant rain. Sheep and cattle perished in their thousands, and disease and famine were rife. Even the arrival in England of the Black Death in 1349 could be traced to a climatic catastrophe seventeen years before in eastern Asia, where probably the worst flood of the millennium drowned seven million people in China and obliterated the natural habitat of the black rat which therefore began a westward migration across Asia towards Europe. During the years of greatest dearth the population in some parts of Europe, including Britain, were reduced to making bread using the ground up bark of birch trees, and the expectation of life declined from 48 years in 1300 to 38 in 1400.

The rest of the fourteenth century delivered widely fluctuating weather with an increase in the frequency of extremes, of heat and cold, and of flood and drought. This excessive variability continued throughout the 1400s as well, and there was a truly exceptional sequence of severe winters in the 1430s – Britain was snowbound throughout six of the seven winters from 1432 to 1438. The eruption of warfare around this time between Scottish clans and the increase in cattle raids into the Central Lowlands and across the border into England can be explained in part by the social stresses induced by a sharply deteriorating climate. There was a markedly warmer period between about 1500 and 1550, and during these decades summers were very much drier than for some two hundred years. Springs and autumns appear to have been quite warm too, but winters remained very variable with long freeze-ups occurring at frequent intervals.

The broad climax of the Little Ice Age began abruptly in the 1550s. Mean summer temperature in the second half of the sixteenth century in England was about 1°C (almost 2°F) below the present level and the mean winter temperature was 1.5° to 2°C (2.7° to 3.6°F) lower than it is now. Severe gales again became more frequent, even during the summers as the Spanish Armada of 1588 found to its cost. Appalling floods once more hit the wide coastal lowlands around the shores of the North Sea, villages were buried by blowing sand as coastal dunes were rearranged (for instance, Udal on North Uist, and Culbin near Nairn), and the most violent storm of the millennium caused immense damage and disruption across southern England on 26 and 27 November 1703. That ferocious gale resulted in the loss of about one-third of England's merchant fleet with between 8000 and 10,000 lives lost, the Eddystone lighthouse was washed away, hundreds of churches were badly damaged, windmills were blown down, and the Bishop of Bath and Wells was killed by falling masonry.

Sea-surface temperature to the north of Scotland was at least 5°C (9°F) lower than now, cold enough to kill off the cod fishery in the Faeroe Islands, along the Norwegian coast, and even in Shetland Island waters. In Scotland itself, famine years were particularly frequent between 1560 and 1625, especially the 1590s, and again in the 1690s when seven of the eight years between 1693 and 1700 brought widespread starvation even in the relatively prosperous east when the oat harvest failed repeatedly. Such extreme climatic pressure led to migration, sometimes forced, from Scotland to Ulster, England, Holland, and eventually also to the American colonies; it also provides an interesting – perhaps telling – background to the political union of Scotland with England in 1707, and to the long-standing tension between the two main communities in Northern Ireland.

In England the main feature of the Little Ice Age was the frequency of long snowy winters, with frost fairs on the Thames in London. Probably the coldest of these, 1683–84, immortalized in R. D. Blackmore's historical novel *Lorna Doone*, was a full degree colder than the twentieth century's most memorable winter, 1962–63, and the ground was frozen to a depth of more than a metre (almost four feet) in parts of southern England. Banks of ice five kilometres (three miles) wide formed around the English Channel coast and in the Thames estuary, and on the Dutch side of the southern North Sea the ice extended 40 kilometres (25 miles) out to sea, halting shipping. Many places in eastern England and the Midlands were snow covered for upwards of 100 days during some of these severe winters; this compares with 60 to 70 days in 1963. Several times Eskimos had ventured as far south as Scotland, presumably due to the great southward extent of Arctic ice at this time; one even paddled his kayak into the estuary of the River Don in Aberdeen, and his vessel and clothing are now preserved in the city's museum.

The improvement which began in the early 1700s was repeatedly scuppered by a marked increase in volcanic activity in many parts of the world. The injection of dust particles and sulphuric acid droplets into the upper atmosphere has a marked cooling effect which may last from one to five years. Memorably severe winters recurred. In 1716 the Thames was frozen so solidly in London that a spring tide lifted the ice bodily 4 metres (13 feet) without interrupting the frost fair, while the winter of 1739–40 was just as long and almost as cold as 1683–84. Indeed 1740 was the coldest year in the entire instrumental record which had begun some 70 years before, the mean Central England Temperature (CET) of 6.8°C (44.2°F) for that year being a full 4°C (7°F) lower than the average figure for the 1990s. By contrast July 1783 was the hottest calendar month on record until quite recently (it has been beaten three times since 1983), while 1784 was so bleak over such a wide area that the Danish government even drew up plans to evacuate the entire population of Iceland. The year 1816 was almost as cold as 1740 and was known as 'the year without a summer' both in Europe and in North America; it followed the tremendous eruption of Tamboro in what is now Indonesia.

The warming trend became far more insistent from 1895 onwards, bringing more frequent warm summers and a marked reduction in the occurrence of severe winters. There was something of a hiccup between 1940 and 1970 but the rise in temperature in all seasons has accelerated since then, particularly after 1988. The last century was about as warm in Britain as the twelfth and thirteenth centuries, but the 1990s, as far can be estimated in the light of present knowledge, constituted the warmest decade of the entire millennium.

Chapter 7: North America since AD 1000

The succession of climatic changes which affected North America was similar to that observed in Europe, although the dates of onset were sometimes different, and the amplitude of fluctuations was also different. There are, of course, no written historical records during the first half of this millennium which means that we know little or nothing about actual weather events during this period, but there is an abundance of archaeological, botanical and zoological evidence to enable us to construct a picture of the general climate.

Warm conditions prevailed over most of the continent until 1200. The coniferous forests of Canada advanced on the tundra, notably in northern Quebec and Labrador, and maize was cultivated by native American tribes in northern Iowa, northern Illinois and southern Wisconsin where it is now too cold to be grown commercially, and in Nebraska and South Dakota where it is presently too dry. The warmth was associated with more abundant and more reliable rainfall over the Great Plains which were more extensively wooded then than at any time since. This benign climate, and the mixed farming that it encouraged, supported a substantial population across the middle of what is now the USA; one town known as Cahokia which was located in southern Illinois had almost 50,000 inhabitants during the twelfth century.

We noted earlier that there was a sharp cooling in Greenland and Iceland around 1200 but that Europe lived on borrowed time for more than a century before temperatures dropped. In complete contrast, the North American climate changed in conjunction with the Arctic and it became abruptly colder and drier during the first decade of the thirteenth century. It is postulated that this change was associated with a strengthening of the westerly flow across the North American continent which reduced the incursion of warm and moist airmasses from the Gulf of Mexico into the Mid-West, the Great Plains, and the Canadian Prairie Provinces. This stronger westerly flow extended across the Atlantic but was diverted well to the north across the European sector, thus maintaining the warmth at the same time as increasing the storminess of winters in the British Isles and adjacent parts of the European continent.

Within four decades the woodlands of middle America had retreated and the Great Plains reverted to grassland, with bison the greatest beneficiary as forest animals disappeared. Native American populations declined and many settlements were abandoned, and further west in Colorado, Wyoming, Nebraska and the Dakotas, the rainfall dwindled so far that the indigenous population was obliged to migrate westward and southward.

Early in the fourteenth century the northern hemisphere circulation appeared to shift; hitherto it had been concentrated on the North American side of the North Pole, but after 1315 the flow became more evenly distributed around the planet, delivering strong westerlies and a marked deterioration to Europe, but allowing greater warmth and more moisture into the southern half of the USA. The earliest settlers in the original colonies, however, experienced a climate still colder than that of the twentieth century, and many perished in the long severe winters which gripped New England, southern Quebec and Ontario during the seventeenth and eighteenth centuries. Thereafter, it seems that the eastern half of North America remained relatively cold until the late nineteenth century, whereas the western half was broadly warmer and moister than it is today. The twentieth century warming trend was detectable over most eastern, central and northern parts of the continent, but it was most prominent in sub-Arctic regions (as in northern Europe). In coastal Greenland, Labrador, Baffin Island, northern Quebec, and around Hudson Bay, the rise in temperature between 1890 and 1940 was three or four times the magnitude of the rise in lower latitudes.

Chapter 8: Mankind's relationship with the weather

The climatic chronology since the last glacial period illustrates quite clearly how human beings have often been at the mercy of long-term climate change, but equally have been able to take advantage of periods of favourable climate. Mankind's relationship with weather (as distinct from climate) has been similarly difficult, although meteorological disasters and triumphs arrive on a very much shorter time-scale than do their climatic counterparts.

In the early days of our history meteorological knowledge was non-existent and human beings were in awe of the weather in its more dramatic and violent guises. So for thousands of years we worshipped weather gods, then seers and sages garnered limited knowledge of meteorological science through keen observation. Geographers refined this process by making lots of measurements which enabled them to recognize and describe patterns, and finally physicists and mathematicians reduced the weather to sets of complicated equations, fed them into computers, and produced pretty accurate weather predictions for six or seven days ahead.

Let us go back a bit. Imagine yourself the village sage several hundreds, perhaps thousands, of years ago, somewhere in Britain. The local folk rely on your wisdom for, among many other things, long-range weather forecasts. These predictions are very important because they enable people to plan for the coming season: how much salted meat, fruit, vegetables and firewood to set in store for the winter; when to start planting crops in the spring; how to order the haymaking and the harvest in summer; and when to gather fruit, nuts, grapes, and so on in the autumn. You are of course no meteorologist even if you knew what the word meant, and you are perfectly aware that your predictive powers are negligible when it comes to the climate of your land. So how do you handle this aspect of your job?

One reason – perhaps the main reason – why you have become the village oracle is the vast store of local knowledge you have accumulated over your long life, a knowledge acquired simply by observing your natural surroundings. A sharp intelligence and a good memory are also very important, an ability to make links and associations is helpful too, and a facility with language which lends gravitas and memorability to your utterances is possibly the most valuable skill of all.

You have noted over the decades that most of your people do not remember from one year to the next how the seasonal progression happens. Their relationship with the weather is at once simplistic and fearful: they expect the normal but fear the catastrophic. Thus in their minds the winters will bring snow and frost, the summers will inflict heat and drought, whereas spring and autumn are transitional seasons providing a gradual change between the two extremes.

Your observational skills and excellent memory tell you otherwise. You know that nearly all winters contain short spells of severe weather, but rarely is a winter icy and snowy throughout. You know that most summers alternate between brief heatwaves and rather longer spells of cooler changeable weather. And you also know that spring and autumn never progress smoothly; rather, the transition between winter and summer (and vice versa) proceeds in fits and starts. Thus, simply by parading a series of truisms about the climate of your village, your reputation as a weather prophet grows apace.

For example, when the cold winds of April have gone and the May sun shines powerfully from a deep blue sky, and everyone thinks that summer has finally arrived, you warn darkly of damaging frosts during the middle of May. Twentieth-century records and statistics show that seven times out of ten you will be right. When the harvest is over, September rains beat against the window shutters, the equinoctial gales rattle the doors and rumble in the chimney breast, and all thoughts turn to winter. But you are able to cheer the populace with tales of summer sunshine in mid-October – latterly referring to St Luke's Little Summer in mid-October and St Martin's Summer in early November, weaving tales around the noble deeds of these saints of old as explanation of these unexpected meteorological gifts. And again, six or seven times out of ten you will get it right.

These ancient sages or prophets built up a substantial store of weather intelligence, and in some early societies they may have gained political and religious power from their skills. In others, perhaps where a hierarchical structure already existed, they were encouraged to share their knowledge for the good of the community, passing it on to subsequent generations of seers and sages. This sharing of weather knowledge was very much the case in European countries, where a large body of weather lore developed over the centuries. When Europeans colonized North America – new lands with new climates – it was imperative that they learnt as much about weather patterns in their adopted surroundings as quickly as possible; thus for such comparatively young countries as the USA and Canada there is an abundance of weather sayings and prognostications, out of all proportion to the length of their history.

Often, an item of weather lore would take the form of a rhyming couplet, all the easier to remember in the days when few people could read or write. Many others were linked to saints' days in the church calendar, making it easier to recall when season-specific indicators were appropriate. In recent times much derision has been heaped upon ancient weather lore by a breed of scientist that has no regard for history, and in truth such simple ground rules could never hope to compete with present-day weather forecasts based on the detailed theoretical knowledge, extensive observational material, and powerful computing techniques that we now have. But those who laugh loudest at old weather sayings – often those with a little learning – take them too literally. For example, those adages linked to saints' days were not meant to be restricted to one particular day in the year: they were designed to indicate a tendency for a particular kind of weather to occur at roughly that time of the year, and the chosen saint's feast-day was merely a useful 'hook' on which to hang the relevant scrap of knowledge.

Arguably the best known piece of weather lore in the English language is a case in point. St Swithin's Day (more correctly, St Swithun), which falls on July 15, marks the interment in Winchester Cathedral of Swithun, who had been archbishop of that English city during the ninth century. Although the old saying takes many forms, the thrust is always the same, as this Scottish version illustrates:

> St Swithin's Day, if ye do rain,
> For forty days it will remain;
> St Swithin's Day, an' ye be fair,
> For forty days 'twill rain nae mair.

Every year in Britain people take great delight in proving St Swithin wrong, and reputable scientists have wasted many pages in journals proving that the saying has never come true since rainfall measurement began in the country over 300 years ago. But they all miss the point. As the eminent English climatologist, Hubert Lamb, put it:

> Now it is a fact, verified by research, that most English summers settle down by about mid-July into some persistent character – either mostly wet, or mostly fine, or repetitive alternations. Indeed this is one of the most noticeable bits of seasonal behaviour in the whole year and is bound up with the behaviour of the atmospheric circulation over most of the northern hemisphere. Moreover, the circulation normally undergoes great changes during June so that the type of weather that is going to settle in for the height of summer often does not become apparent until nearly July 15. The next pronounced shake-up of the general pattern of winds and storm tracks over the hemisphere is normally around August 24 so that there is indeed a tendency for our spells of high summer to end about that time.

Thus the author(s) of the Swithin rhyme had observed a marked tendency for the weather to settle into a pattern for five or six weeks from about the middle of July onwards, long before Lamb and his contemporaries had the scientific information to explain why. We should doff our caps to them.

Perhaps more than any other month November is replete with old country weather lore. People feared the winter because it brought not only damp and discomfort, but also food shortages, and often illness and death too. A mild winter was regarded as particularly unhealthy because it was widely held that a spell of sharp frosts was needed to kill off the diseases which flourished in the moist winter air. However, a prolonged severe winter was just as bad, with many dying from hypothermia and malnutrition.

Seasonal indicators which developed independently in several European countries include:

> If November ice will bear a duck
> There'll be nothing after but slush and muck.

and

> A warm Christmas, a cold Easter;
> A green Christmas, a white Easter.

and

> He who shears his sheep before St Saviour's Day
> Loves his wool more than his sheep.

The feast-day of St Saviour falls on May 13, and along with St Mamertus' Day on May 11 and St Pancras' Day on May 12 forms the festival of the Ice Saints. Mid-May is still regarded as a period when sudden sharp frosts can cause untold damage to young vegetables and fruit blossom.

From the eastern states of the USA we also have:

> Heavy September rain brings drought in its tail.

and

> When birds and badgers are fat in October, expect a hard winter.

and

> A heavy November snow will last till April.

Groundhog Day, now world famous following the Hollywood film, falls on February 2 and is a direct steal from the European Candlemas Day which has a myriad old sayings linked to it. Like the American version, they all warn of the likelihood of renewed wintry weather round the corner if early February produces mild sunny conditions – sound advice.

Apart from the seasonal omens, there is a large mass of weather lore which addresses day-to-day weather changes, mainly by keen observation of changes in the sky and the wind. Some of these are only appropriate to the mid-latitude zones of westerly winds, and only work well towards the western seaboards. But this does mean that sayings originating in England maybe a thousand years ago will still work in Seattle or Vancouver or Wellington, and probably also in Cape Town and Hobart. Thus we have the well-known

> Red sky at night,
> Shepherd's delight;
> Red sky in the morning,
> Shepherd's warning.

and

> Rain before seven,
> Fine by eleven.

and

> Rain long foretold, long last;
> Short notice, soon past.

All these can be explained in meteorological terms when depressions and their associated frontal systems sweep in from the ocean at regular intervals, separated by ridges of high pressure.

However, it is difficult to find any reason in some of the old rhymes which attribute forecasting skills to animals and plants, especially as there are a number which simply contradict each other. The last century or two have given us little fresh weather lore, but instead a few examples of irony and parody:

> The rain it raineth every day
> Upon the just and unjust fella,
> But mostly on the just, because
> The unjust hath the just's umbrella.

and

> Dirty days have September, April, June and November.
> From January up to May
> The rain it raineth every day;
> All the rest have thirty-one
> Without a blessed gleam of sun;
> And if any of them had two and thirty
> They'd be just as wet and twice as dirty.

Chapter 9: The scientists take over

The UK's Meteorological Office is moving to Exeter, Devon, in 2003. Hitherto it resided in Bracknell, Berkshire, in a building functional rather than inspirational. Over the years it has overflowed into a number of nearby annexes, but there is very little that distinguishes it to the casual observer as the headquarters of one of the leading national meteorological services in the world. In the 1960s and 1970s the radar antenna on the roof may have given a clue to the occupant's corporate identity, but these days such high-tech excrescences are two-a-penny and you really have to read the comparatively modest nameplate outside the front door to be sure where you are.

How unlike its distant predecessor, the Tower of the Winds, constructed almost 2000 years before the Bracknell building, and which remains one of Athens' more notable architectural survivals and tourist attractions. To be fair, the octagonal tower, completed in approximately 35 BC, was not built as an observatory; rather, it was the Big Ben of its day. Andronikus of Kyrrhos was the architect, and what he designed was a giant water-clock – the Horologium – which was driven by a small aqueduct entering the back of the tower. The water-clock, of which little remains, was supplemented by a series of sundials which still function. The roof was topped by a wind-vane fashioned in the likeness of the sea god, Triton, one of the earliest working wind-vanes that we are aware of.

Each of the eight marble faces on the outside of the tower has a male figure carved into it representing the personification of the wind blowing from that particular direction. Thus the north-facing wall shows Boreas, the north wind, while that facing south is decorated with a figure of Notos, the south wind. The other minor wind-gods displayed are Kaikias (north-east), Apheliotes (east), Euros (south-east), Lips (south-west), Zephyros (west), and Skyron (north-west). All these figures are reproduced on the tower of the Radcliffe Observatory in Oxford, and also in the library of the Blue Hill Observatory near Boston in Massachussetts. The Tower of the Winds itself is used as the logo of the Royal Meteorological Society. Each of these wind gods has a character of his own, supposedly representing the weather associated with his particular wind direction. Thus Zephyros is a lightly clad youth bearing flowers, whereas Boreas is an old man wearing thick clothes and carrying a conch shell to indicate the howling of the wind. Apheliotes is a young man bearing fruit and corn, and Notos is a virile youth carrying an inverted pot from which water is pouring.

The Ancient Greeks were probably the first to apply scientific thought to weather phenomena. In the fourth century BC Aristotle wrote a treatise on the subject entitled *Meteorologica*. He clearly understood some aspects of the hydrological cycle whereby water evaporates from seas, rivers and lakes into the atmosphere, whence it condenses into rain or snow which falls back to the ground:

> The finest and sweetest water is every day carried up and is dissolved into vapour and rises to the upper region, where it is condensed again by the cold and so returns to the Earth ... so the sea will never dry up, for before that can happen the water that has gone up beforehand will return to it.

Some decades later Theophrastus, who studied under Aristotle, wrote an essay on 'Weather Signs' which incorporated further scientific reasoning, including a crude understanding of air-masses and orographic rainfall processes. However, many of their other theories, such as the cause of thunder and lightning, were wide of the mark. Natural philosophers such as Ptolemy and Hero of Alexandria, and Philo of Byzantium, constructed the first rudimentary instruments which showed how temperature and pressure changed with time.

After the Greeks little advance in meteorological science was made in European civilization until the Renaissance. Our attention now switches to sixteenth-century Italy, where scientific innovation flourished thanks to the patronage of various princes and dukes. The astronomer Galileo (Galilei) devised some rather primitive equipment for observing fluctuations of temperature and atmospheric pressure, but it was well into the

the seventeenth century before the sealed mercury-in-glass thermometer and the mercury barometer – instruments that we would recognise today – were invented. Interest in meteorological measurement quickly spread to France, England, Germany, Holland, Denmark and Sweden, and developments in physics during the seventeenth and eighteenth centuries led to a greater understanding of fundamental atmospheric processes. This period also saw the first systematic daily weather readings instituted at a number of observatories across Europe, and by the 1790s the first networks were established which brought together the observations over a wide area for analysis at a central location. The first of these was developed by the Societas Meteorologica Palatine in Germany and the Royal Society in England. Nevertheless general text books covering weather and climate had scarcely advanced from the Aristotelean view even as late as 1850.

The more perceptive nineteenth-century meteorologists quickly realized that a knowledge of conditions in the upper atmosphere was imperative to understanding how the weather worked. Following the Montgolfier brothers' first hot-air balloon flights in 1783, it was not long before scientists started taking meteorological instruments ever higher into the atmosphere. The French chemist, L-J Gay-Lussac, discovered that most of the water vapour in the atmosphere is confined to the lowest layers, with precious little moisture above eight kilometres (five miles) aloft. John Wise, an American, was surprised by the strength of updraughts in cumulo-nimbus clouds when he flew a balloon into a thunderstorm in 1843. And in 1862 the veteran English meteorologist, James Glaisher, nearly lost his life when his expedition took him to an altitude of nearly ten kilometres (six miles).

The invention of the electric telegraph in the middle of the nineteenth century gave a tremendous impetus to meteorological research. For the first time weather observations could be performed, transmitted, and analysed at a central location within an hour or two. Experts were now able to construct a much more complete picture of weather patterns in real time, and at regular intervals throughout the day.

In 1854 Admiral Robert FitzRoy (already famous for having taken Charles Darwin on a voyage around the coastline of South America in the *Beagle*) was appointed superintendent of the newly established Meteorological Department of the Board of Trade – the earliest incarnation of the UK Meteorological Office. FitzRoy gathered simultaneous weather reports from over forty weather stations around the British Isles and neighbouring parts of Europe, and from this information drew maps showing details of the weather at a given time. These were the first 'synoptic charts'. In 1861 he initiated a storm warning service for shipping, and in 1864 he started issuing general forecasts for the country which were published in *The Times*, much to the consternation of some of his colleagues and superiors. The predictions were not very good, they were often derided by the public, and before long the editor of *The Times* decided to discontinue the service, shortly before FitzRoy took his own life. The gale warnings for coastal waters, however, continue to this day. The Americans also began forecasting, and in 1871 boasted a 75 per cent accuracy rate in their predictions of rainfall and temperature – an inflated claim.

The development of air travel in the early 1900s accelerated the development of the science. The early aviators in the Great War were keen to learn about meteorology, and their commercial successors in the 1920s and 1930s were voracious consumers of weather information and forecasts. Without a doubt aviation was the most weather-sensitive activity of the twentieth century. Not only did they need an accurate description of wind and cloud conditions during their flights, but they also needed predictions of the weather at their prospective destinations and also at alternative destinations should they be diverted. Aviators were also contributors. They carried on-board equipment which measured temperature, humidity, wind and air pressure, and these details were radioed back to the meteorological offices. Systematic upper-air observations began in the 1940s following the invention of the radiosonde – a collection of sensors together with a radio transmitter attached to a balloon – which were released twice daily from a network of stations across the world.

In the last fifty years technological advances have accelerated. The advent of satellite images, growing computer power, satellite sensing of temperature and wind, and radar measurement of rainfall intensity and wind speed, have all helped to improve the accuracy of our forecasts. Thus we now have huge amounts of information about the atmosphere, updated at frequent intervals during the day. Computers are just about big enough to handle it all, and the theoretical scientific knowledge is available to predict what will happen with reasonable accuracy up to six days ahead. Forecasts still occasionally go wrong, but that just shows what an enormously complex machine the atmosphere really is.

Chapter 10: Weather events in history

We have seen how mankind has coped with climatic changes extending over centuries. Weather events, of course, occur on a very much shorter timescale and therefore can have a much more dramatic impact. As our society becomes more and more complex, those events can impinge on our everyday lives in a devastating fashion. An ice-storm in Canada in January 1998 left hundreds of thousands of people without power for several weeks in mid-winter; record-breaking floods in the UK in autumn 2000 wrecked thousands of homes and disrupted transport for months; brush fires in New South Wales in January 2002 left hundreds homeless.

Occasionally, the weather may have repercussions which reverberate down the corridors of time. Here are some examples.

The inhabitants of northwest Europe do not experience the extraordinary violence of a mature tropical hurricane, or the catastrophic power of a fully grown tornado. Nor do they have to endure the extremes of cold, heat and drought that are part of the fabric of the global climate, and which other nations may regard as 'normal', notwithstanding the periodic loss of life that may occur. But one of the greatest meteorological disasters of the twentieth century happened in this part of the world when a ferocious northerly gale drove a massive storm surge from the Atlantic into the almost enclosed waters of the North Sea on the night of January 31 to February 1, 1953. The increasing shallowness and narrowness of the southern North Sea mean that water piles ever higher as such a storm surge penetrates past the 55th parallel, resulting in disastrous floods in the Netherlands, eastern England, north-western Germany, and to a lesser extent in western Denmark, Belgium, and the extreme north of France. Some 2000 people were drowned and 100,000 had to leave their homes, about 8000 square kilometres (over 3000 square miles) of land were inundated, and livestock losses were put at almost 300,000. Total insurance losses at 2002 prices were in excess of £20 billion ($30 billion). The Dutch had to race against time to seal the hundred-odd breaches in their dikes before the following winter.

This was the latest of a long sequence of North Sea floods which can be traced back to the flooding of the North Sea and isolation of Britain from Europe shortly after the end of the last glacial period. Indeed, some of the biggest weather-related death tolls through history have occurred in those vulnerable and heavily populated lowlands bordering the North Sea. The climate of the late Middle Ages appears to have been particularly vigorous, with the frequency of gales higher than at present, and the limited records from those times provide tantalizing glimpses of the extent of these human tragedies – not just the huge loss of life, but also the permanent loss of vast tracts of land to the sea. The earliest useful estimates of casualties concern the floods of January 1281 in the Netherlands and December 1287 in Germany; between 30,000 and 80,000 people are estimated to have been drowned on both occasions. Those of January 1362 and November 1421 may each have taken the lives of 100,000.

The 1362 storm is known as the *Grote Mandrenke*, the *Grosse Männdrenke*, or the *Great Drowning* and the number of people drowned on that occasion may have amounted to half the population of southern Denmark and north-west Germany. This was a formidable disaster coming just a few years after the ravages of the Black Death. England escaped the worst of the flood, but the high winds caused some loss of life, and many church towers were demolished in London, Norwich and Bury St Edmunds.

The flood of November 1–2, 1570, known as *the All Saints' Flood*, was probably the worst of the lot in terms of lives lost, as the heavily urbanized lowlands of the Netherlands were the worst hit and the cities of Amsterdam, Rotterdam and Dordrecht were totally inundated. Contemporary chroniclers estimated that between 100,000 and 400,000 were drowned. Such phenomenal losses clearly concentrated the minds of the governing classes, especially in the Netherlands where a 400-year-long battle against the sea was begun, dikes and barriers were raised and strengthened, drainage channels constructed and improved, and land was reclaimed from the waters. Meteorologically speaking, the gale and storm surge of January 1825 were even more violent than those of 1570 and 1362, there was widespread gale damage and considerable flooding, but the death toll was under 1000 thanks to new sea defences. North Sea storms have not stopped happening, and this catalogue of death and destruction emphasizes how much modern western European society takes for granted the sea-defences along North Sea coasts.

Afflavit Deus et dissipati sunt was the inscription on the medal that Elizabeth I awarded to naval combatants in the sea battle against the Spanish in the summer of 1588: 'God blew, and they were dispersed'.

The story that the Spanish Armada was vanquished by exceptional summer storms and not by the Royal Navy was long encouraged by the English political establishment and repeated by some historians. Such an idea is not as perverse as might appear at first glance: it suited the English that God should appear to be on their side. On the isolated occasion – July 19 – when the weather favoured the Spanish, a strong southerly wind bottled the English fleet in Plymouth Sound. Rather than countenance any suggestion that God might favour their opponents, however briefly, the English devised the myth that their ships remained in port while Drake finished his game of bowls. Political spin, it would be called nowadays.

Modern historical commentators have a more balanced view of the defeat of the Armada. It is now generally accepted that the English had done for the Spaniards before any serious meteorological intervention. The Tudor historical expert, Geoffrey Elton, summarised it thus: '... [the Spanish] were defeated by better ships, better seamanship, a needlessly rigid plan, and at the last the weather.' None the less there was disappointment even at the very highest level that so few ships were captured.

One aspect of the historical accounts of this battle is particularly striking to a meteorologist. Conventional historians occasionally make a cursory reference to the weather when it impinges on historical events, but they rarely do more than scratch the meteorological surface. For instance, Elton talks of 'a contrary wind', 'a rain squall from the north-east', and 'the wind shifted back to the south-west'. He does not investigate how unusual the storms of July and August 1588 were, nor does he make reference to the painstaking work of climate historians who have pieced together the sequence of weather events during the crucial weeks from ships' log-books and from daily weather observations from a handful of sites in England and adjacent parts of the continent.

Research workers at the University of East Anglia deduced that the scattering of the Spanish fleet in the North Sea during the first week of August was due to two intense Atlantic depressions which tracked north-eastwards across northern Britain and thence to Scandinavia. The complex sequence of wind directions logged by both Spanish and English ships might at first sight appear almost random, but they fit exactly into the pattern suggested by the meteorological reconstruction. In particular, the violent north-westerly gale in the wake of the second depression would have been exactly the sort of event to cause the severe losses suffered by the Armada as it limped home around the Atlantic coasts of Scotland and Ireland. According to the scientists at the University of East Anglia these storms were of an intensity outside the range of summer depressions during the twentieth century, but may have been quite common during the cooler, windier summers of the Little Ice Age.

Almost 200 years later the weather foresook the British cause and fought on the side of the Americans at several crucial points during the War of Independence. On March 5, 1776, a severe south-easterly gale scuppered General Howe's plan to launch an amphibious assault on George Washington's newly-gained positions on Dorchester Heights, just south of Boston, by preventing British ships from entering the harbour. The delay enabled the Americans to entrench, bringing up their artillery and fresh supplies. From here Washington was able to threaten the British troops in the town, and before long Howe decided to retreat from Boston altogether. At the Battle of Long Island, five months later, the British held the upper hand but a sequence of weather events enabled the Americans to retreat from an impossible position. Firstly, on August 27, wind and tide prevented the attackers from entering the East River whence they would have captured the American positions on Manhattan Island, and, in the words of the war historian Christopher Ward, 'the revolution might have ended then and there.' General Washington ordered an evacuation for the night of August 29 and expected heavy losses, but a thick fog, rare in summer, developed during the early hours of the morning, and Washington's forces escaped unseen.

Following defeat at Trenton just after Christmas 1776, British troops under General Cornwallis launched a counter-attack at Princeton as the new year dawned. A sudden thaw helped, turning the hard-frozen roads into deep quagmires which slowed the advance of American reinforcements. Late on January 2 Cornwallis thought he had won, and decided to wait until first light the next day to surround the enemy and capture his opposite number; but once again the weather changed, the temperature dropped dramatically overnight, and the roads re-froze, enabling Washington to outflank the British, inflict severe losses on their rearguard, and then to march off to the north before midday. After midday on January 3 the thaw resumed, preventing the British from following.

The weather intervened on the American side one last time during the Siege of Yorktown which preceded the final British surrender in the autumn of 1780. The King's troops had been encircled by the Americans and their French allies for almost three weeks when General Cornwallis planned an audacious breakout across the York River, from where he hoped to be able to join up with his countrymen, who still controlled large areas to the north. The plan, scheduled for the night of October 16–17, involved sixteen boats making three round trips each, and needed light winds and still waters to be successful. Halfway through, a fierce north-westerly squall accompanied by rain and hail struck, driving all the boats downstream and into the arms of the enemy. Cornwallis's troops were now divided by the river, he found himself in an unacceptably vulnerable position, and raised the white flag later that morning. Terms were agreed and the surrender was formalised on the 19th.

Cases can be made for decisive meteorological intervention at several other key events in history. Mention has already been made of the sequence of severe winters and harvest failures in Scotland at the end of the seventeenth century which preceded the Act of Union between Scotland and England in 1707. Some historians argue that Napoleon was defeated as much by heavy rain and mud as by Wellington. And the excessive rains of 1916 and 1917 – especially during the summer of the latter year – may have extended the stalemate in Flanders during the Great War. Earlier in 1917 the Russian revolution followed hard on the heels of a particularly severe and protracted winter in eastern Europe, the coldest in St Petersburg since 1895. The great Russian winter also had a considerable impact on the Second World War, as the extreme severity of 1941–42 stymied Hitler's attempts to take Moscow. More recently, the dust-laden atmosphere of central Iran may have contributed to the failure of the American operation to rescue the embassy hostages in April 1981; a combination of poor visibility and human error led to a humiliating defeat with considerable losses of equipment and personnel.

The weather was undoubtedly important during the Gulf War in early 1991, although it probably did not have any material impact on the outcome. The Iraqis, under Saddam Hussein, invaded Kuwait on August 3, 1990, but it was over five months before the anti-Iraqi alliance was in a position to launch a counter-offensive to liberate the small oil-rich Gulf state. This was probably the first war fought on our television screens, and from those images it was clear that meteorological conditions were often important, and that weather forecasters undertook a vital if unsung role.

We naturally think of deserts as places where it never rains. This is, of course, a gross generalization, and it is certainly not true of the Arabian peninsula during the winter season. Iraq and Saudi Arabia are protected from icy blasts from Siberia only by the mountain ranges of Iran and Turkey, while winter depressions from the Mediterranean bring cloud and rain from the north-west at intervals. In the war zone, daytime temperatures were typically between 15° and 20°C (59–68°F), dropping to between 5° and 10°C (41–50°F) at night. Early mornings were quite humid – fog was an occasional problem – and the average cloud-cover was around 60 per cent. In other words, the sky was more often cloudy than not, disrupting reconnaissance missions. Early 1991 was wetter than average, and rain fell in the region a couple of times a week, occasionally heavily and accompanied by hail and thunder. When it rains the dust of the desert turns quickly to mud, while during the drier interludes a strong wind will easily lift the dust and sand from the desert floor, resulting in sandstorms.

It was very interesting during the period of hostilities to compare the pictures we were getting on television news bulletins with what could be deduced from weather-satellite images. Over and over again there were mismatches. For example, the pictures of news correspondents earnestly explaining the problems of transporting military equipment across a sea of mud through torrential downpours and under a lowering sky appeared to coincide with satellite pictures showing a thousand-mile-wide cloud-free swathe covering the whole of the Arabian peninsula. The conclusion was that the so-called 'actuality' from the war zone itself was in fact two days old, enabling the military authorities to control the news output throughout the war, whatever the radio and television people based in Riyadh would have liked us to believe. On one occasion a meteorologist in London described this delay in a radio broadcast, only to be politely told off by the Ministry of Defence, who described his comments as 'unhelpful'.

There was much concern at various points during and after the Gulf War about the possible impact of the pollution from oil fires on the global climate. This largely took the form of irresponsible and sensationalist stories in the print and broadcast media which were really no more than axe-grinding and pot-stirring from environmentalist pressure groups and tabloid editors – equally meteorologically illiterate – who unusually found themselves in the same bed. Had they sought scientific advice they would have learnt that the atmosphere in the Middle East is what meteorologists call 'stable'. This means that air tends to sink most of the time, especially in the middle and upper parts of the troposphere. When the oil refinery in the Saudi Arabian town of Khafji was bombed by the Iraqis early in the war, we saw an impressive plume of smoke on the TV pictures, but it was also clear from these images that the smoke was rising about 600 metres (2000 feet) into the atmosphere and then it just stretched out horizontally; it could climb no higher because of the descending air currents. Similarly, when the Kuwaiti oil wells were fired by the retreating Iraqi troops, nearly all the pollution remained in the lowest three kilometres (two miles) of the atmosphere. It was thus prevented from reaching the stratosphere, which is what would have had to happen if the smoke and soot were to be transported around the northern hemisphere. The pollution was thus concentrated in the local area, and there was never any real chance that the Indian monsoon would be affected as some feared, let alone weather patterns in other parts of the world.

SECTION TWO

Weather at the turn of the millennium

Chapter 11: Monitoring the weather, 1960–2000

The last forty years of the twentieth century saw a revolution in the way the atmosphere is monitored and in the way the weather is predicted. Computers arrived much earlier in the weather business than in most other sectors of the economy, and they were already having a considerable impact on forecasting by the mid-1960s. Most departments of a meteorological office had computers installed by the late-1970s although – as elsewhere – the arrival of the digital age in the 1990s brought a fresh surge of computer development.

In 1960 a typical observing station comprised four thermometers in a Stevenson screen, a simple rain gauge, an instrument for measuring sunshine duration, and some clever mechanical devices for monitoring rainfall intensity, wind direction and wind speed. Depending on the importance of the site, weather observations would be carried out half-hourly, hourly, three-hourly, six-hourly, or once a day. A few stations would also release radiosonde equipment, tracked by radio, twice a day to measure temperature, humidity and wind in the upper atmosphere, but there were not many of these because the hardware was expensive. The observer would then transform the readings into a line of code comprising several groups of five digits, then send it by telephone or teleprinter to a regional collecting centre. Within an hour, all the observations in the country were broadcast by teleprinter operating at a speed of five characters per second.

By 2000 the Stevenson screen and its thermometers can still be found at most observing sites, although the manually read instruments are gradually being replaced by electronic stations which log data continuously throughout the day. Radiosondes are still used too, but these traditional tools are only one small part of a vast array of data-gathering hardware which monitor the atmosphere both near the ground and high above it all the time. These include satellites, rainfall radar, Doppler radar for measuring winds, radiometers, meteorological buoys, drop-sondes released from aircraft, sky cameras, lightning detectors, and a variety of remote-sensing equipment. Someone using the extraordinary mass of data could become snowed under very rapidly, so most of the incoming information is sorted electronically. However, surface and upper-air weather observations are still transmitted on a global telecommunications system, by computer rather than teleprinter now, and at a speed of over 10,000 characters per second over a modem link – very much faster even than over a hard-wire link.

A typical forecasting office in 1960 was peopled by two 'castes' – forecasters and assistants. The job of an assistant was to carry out the regular weather observations, and to plot all the data from around the country and around the world, both at the surface and in the upper air, on charts. Some would become exceptionally proficient at chart plotting, using a double nib-bed pen with black and red inks, completing each station report in under five seconds. Thus a chart with observations from over 500 stations plotted on it could be completed in less than an hour. It was then passed on to one of the forecasters whose first job was to analyse the chart, determining where fronts were, drawing isobars, and highlighting zones of significant weather.

Everyone worked at large sloping benches, and the forecaster's main tools were a soft lead pencil, an eraser, and some coloured crayons. The senior forecaster would consider all the latest charts – known as analyses or baratics – and using rules and guidelines developed by his predecessors over many years, leavened by his own experience, a measure of gut-feeling, and consultation with his colleagues, he would construct a sequence of forecast charts – prognoses or prebaratics – covering the next 24 or 48 hours. Anything beyond 48 hours was often little better than guesswork. The forecasters would then use these forecast charts as a basis for the variety of forecasts they were required to issue.

The office would be noisy with people having to shout to be heard above the endless clatter of teleprinters churning out miles and miles of paper; indeed, there would be paper everywhere, charts on benches and walls, observations on clipboards, forecast forms scattered around, piles of new teleprinter rolls overflowing from cupboards, and used teleprinter reports discarded in corners.

An assistant in 1960 could well have ended up a senior forecaster by 2000. His office is very different. Assistants have vanished, apart from one or two office juniors, and there are fewer forecasters too. They work at ordinary desks with at least two computer terminals, one for displaying data and one for creating his forecasts – this really is a paperless office. It is also a quiet one, with an almost inaudible hum of hard disks and air-conditioning units, the latter required to disperse the heat generated by all the electronic equipment. A new breed has now arrived, the computer support staff, and there will probably be an office manager whose job it is to ensure the forecaster is not unnecessarily interrupted from his routine.

Nearly all the raw forecast material is now produced at the nation's central forecasting office, and comprises computer output overseen and occasionally adjusted by a human forecaster – the so-called 'man-machine mix'. At the outstation the forecaster's job is simply to interpret that output, incorporate local knowledge, and present his forecasts according to the customer's requirements. More and more of this is being automated, and the need for the human forecaster on site is diminishing. Not very long ago all pilots, from captains of jumbo-jets to weekend enthusiasts, would get a meteorological briefing from an onsite forecaster; nowadays, pilots along with mariners and offshore oil-industry personnel, are encouraged to 'self brief'.

There is one other big change from 1960. Forty years ago most weather forecasters in most countries around the world were civil servants or military personnel, working for the national meteorological service, and the bulk of the forecasting work was for civil and military aviation. Things had already changed in the USA, where the commercial sector expanded rapidly after the end of the Second World War. In other countries the numbers employed in meteorology reached a high-water mark in the 1960s and early 1970s; progressive computerization of chart plotting and forecast production together with growing financial exigencies led to a contraction in employment, especially after the oil crisis of 1973 and again after the recession of the early 1980s. Private forecasting companies received a fillip in countries with an offshore oil industry in the 1970s, and a slow and fitful growth throughout the 1980s and early 1990s saw a diversification in the private sector into radio and television, newspapers, insurance, marine operations, and finally online services. Except in the USA, the greatest employer of weather forecasters is still the national meteorological service, although a few of these, that of New Zealand for example, have themselves been privatised. But the private sector is now offering a significant number of employment opportunities for meteorologists in the United Kingdom, the Netherlands, Germany, France, Canada, and many other countries, as well as the USA.

Chapter 12: Instruments and observers

Scientists in ancient Greece made the first faltering steps towards designing instruments to study the behaviour of the air around them, although they were more interested in demonstrating a variety of physical principles rather than in measuring anything. Nevertheless their ideas formed the basis for the invention of the thermometer, the barometer and the hygrometer – those familiar instruments which quantify the temperature, the pressure and the humidity of the air around us.

How warm is it?

After the Greeks no significant advances took place until the Renaissance, when scientific thought began to free itself from the dead hand of fundamentalist religion. Galileo, the famous Italian astronomer who lived from 1564 to 1642, conducted some experiments to show that air expanded when it was heated and contracted when it cooled down again. His first instrument – hardly yet a thermometer – used coloured water to illustrate the change in temperature, but the glass tube containing the water was not sealed and therefore changing air pressure as well as changing temperature had an effect on the level of fluid. As a consequence it could only indicate qualitatively how warm it was.

The first sealed thermometer dates back to about 1640. It is sometimes credited to Ferdinand, Duke of Tuscany, although it was more likely to have been invented under his patronage by a court scientist. The court resided in Florence, and this first true thermometer is usually referred to as the 'Florentine thermometer'. Alcohol, distilled from wine, was used as the indicating fluid, and a rudimentary scale was devised using the blood temperature of animals as its upper point and the temperature of snow (an occasional visitor to Florence in most winters) as its lower point.

Alcohol has one big disadvantage; it boils at 78.5°C (175°F), so it could not be used in scientific experiments involving temperatures higher than this. Water, of course, has a problem at moderately low temperatures, and the freezing process usually shatters the glass tube. The familiar mercury-in-glass thermometer was devised by Florentine scientists in 1657. Mercury is an ideal substance for thermometry because it expands and contracts rapidly as the temperature changes – it is said to have a high coefficient of expansion – and it remains in liquid form between −39°C (−38°F) and 357°C (673°F). One problem encountered with mercury thermometers, even to the present day, is that minor imperfections or tiny particles of foreign matter on the inside of the tube will result in small droplets of mercury sticking to the glass, thus rendering the temperature reading inaccurate. The manufacture of these thermometers, therefore, has always been a high-precision task.

In England, the Royal Society entrusted the development of thermometer design and manufacture to one of its Fellows, the Irish physicist Robert Boyle. He had spent some time studying in Florence and around 1664 he produced a sealed alcohol thermometer, and in the UK alcohol and mercury were both widely used until the beginning of the twentieth century. At about the same time, in 1665, the English physicist Robert Hooke, a colleague and a friend of Boyle, invented a scale for use on their thermometers which had its only fixed point at the melting point of ice. It is believed that this may have been the earliest such scale with its zero at this point, predating the one which bears Celsius' name by some 80 years. There were several schools of thought as to what the upper point should be: some favoured blood temperature while others chose the boiling point of water, or the temperature of the hottest day of the year.

During the eighteenth century the Fahrenheit and Celsius (or centigrade) temperature scales saw off all competitors. For meteorological purposes the degree Celsius remains the official worldwide standard to this day, even though the SI unit of temperature, used by physicists and other scientists, is now the Kelvin. One Kelvin (not one degree Kelvin) is equal to one degree Celsius, but the zero mark is at minus 273.16°C – often called Absolute Zero. Indeed the scale itself used to be called the Absolute scale. The zero point depends on the fact that the volume of a gas decreases as the temperature decreases, and all gases eventually vanish if their temperature is reduced to Absolute Zero.

For more than one hundred years after the first sealed thermometer was invented in 1657 a dizzying variety of temperature scales were used, and the calibration of thermometers was, to say the least, haphazard. But by the middle of the eighteenth century three of these scales had achieved take-off in terms of popularity – those bearing the names of Fahrenheit, Celsius and Réaumur.

Many parts of the English-speaking world have a curious attraction to the temperature scale developed by Fahrenheit. In Britain, especially, this support might almost be called jingoistic, a totem for the anti-Europeanists, were it not for the fact that Daniel Gabriel Fahrenheit was an arch-European. Of German parentage he was born in 1686 in the free city of Danzig (now the Polish port of Gdansk), conducted most of his work in the Netherlands, and reported his findings to the Royal Society in London. It is not generally known that Fahrenheit had invented an earlier temperature scale which he used on thermometers that he manufactured commercially before 1708. This scale extended from 90 degrees, which he described as 'the greatest observed warmth', to minus 90, which was 'the greatest observed cold'.

It is now believed that Fahrenheit was inspired to produce a new scale following discussions with the Danish mathematician Olaf Römer around 1709, although documentary evidence of these meetings is lacking. Fahrenheit's new scale, the one that now bears his name, had three fixed points: zero was specified as the temperature of a mixture of water, ice, and ammonium chloride; 32 was the temperature of a mixture of ice and water; and 96 was defined as the temperature of a healthy person's *axilla* (armpit). This scale rapidly gained popularity in northern Europe, especially in Britain, and Fahrenheit was elected to the Royal Society to which he communicated three scientific papers explaining its derivation.

Anders Celsius was a Swedish physicist and astronomer who lived from 1701 to 1744, and in 1742 he published a scientific paper describing a thermometer with the freezing point at 100 degrees and the boiling point at zero. This was not the first 'centigrade' or 'centesimal' scale to be used in Europe, and as a young scientist Celsius would almost certainly have used thermometers devised by his compatriot, Palmburg. Palmburg's scale had a zero reading representing the temperature of the coldest day of the year, and 100 as the temperature of the hottest day of the year. Celsius probably chose to reverse the idea of having a high figure to represent hot conditions, and vice versa, in order to distance his efforts from Palmburg's – in other words to make his own work appear to be original and not leave himself open to the accusation of plagiarism.

The concept that a high temperature represents heat is so ingrained in our modern psyche that it is difficult for us to appreciate that this was not so 250 years ago. The reversal of Celsius' original scale is usually attributed to the Swedish botanist Linnaeus, but there is no published evidence for this. Another Swede, Marten Strömer, certainly used the reversed scale (with zero at freezing point) before Linnaeus did. Nor is there any corroboration for claims – found in some modern reference books – that Linnaeus used such a thermometer before 1740.

More interesting, though, is evidence that a French scientist, Jean-Pierre Christin, independently developed the present centigrade scale around 1740 or just after. Christin was, it seems, a modest man who lived in the shadow of his much more celebrated compatriot, Réaumur. Christin found it difficult to publish his work, while his fellow Frenchman may even have discouraged competition. René-Antoine Ferchault de Réaumur invented the scale which bears his name in 1730; it had freezing point at zero and boiling point at 80, and Réaumur thermometers can still be found in some countries in southern Europe, notably Romania and France, and also in former French colonies in Africa.

As light as air

We now know that barometric pressure – that is, the downward pressure exerted by the air above us – is one of the three most important characteristics of the atmosphere that we can measure. The others are temperature and humidity. However, air pressure is something we are unaware of in most circumstances. It is therefore surprising that the first barometer was invented within a few years of the first thermometer. It is perhaps less surprising that Galileo was involved in both inventions. Galileo had spent some time working on the efficiency of water pumps and he became aware that, however good the pump was, it was impossible to raise water more than 10 metres (33 feet) above the original water level.

Galileo failed to discover exactly why these water pumps failed, but within a couple of years of his death one of his pupils had cracked the problem. His name was Evangelista Torricelli, and the apparatus he devised to demonstrate that air exerted a downward pressure was actually a prototype barometer. He discovered that air pressure will support mercury in a glass tube to a height of some 76 centimetres (30 inches), and he deduced that the difference between the 76 centimetres of mercury and the 10 metres of water was due to the different densities of the two fluids.

For a long time barometric pressure was measured in the English speaking world in 'inches of mercury', and eventually in other European countries in centimetres or millimetres. But air actually exerts a downward force on the Earth's surface, and pressure is officially defined as 'force per unit area'. Thus by the early twentieth century it was accepted that it would be more accurate to measure barometric pressure in these terms, so the 'millibar' was introduced. This in turn was re-named 'hectopascal' – one hundred pascals – after the seventeenth-century French scientist Blaise Pascal, and a pascal is defined as a pressure of one newton per square metre. Despite this change it will probably be several decades before weather forecasters finally relinquish their beloved millibars.

Every drop of rain that falls...

Measuring rain probably appeared to be a simple operation to early weather enthusiasts: all you had to do was place a container out in the open, then measure the depth of water in it at regular intervals – say once a month. But it was quickly realized that, especially in sunny climates and especially in summer, a large proportion of the rainwater would evaporate when the sun shone and the wind blew. Additionally, sometimes in heavy downpours drops of water splashed into the containers, and occasionally some water splashed out as well. It also became apparent that daily values of rainfall deserved study, and very often these were exceedingly small and therefore very difficult to measure.

Rain gauges therefore soon evolved into double sided cylinders to minimize evaporation, surmounted by a funnel with a substantial vertical lip to reduce splashing. The rain gauge was, and still is, usually made of copper. Various sizes of funnel were tried, but those with a diameter of either 12.7 centimetres (5 inches) or 20.3 centimetres (8 inches) became the standard. If the water caught in these gauges was poured into a much narrower measuring glass, small amounts of rain could be measured more accurately. In the early days of rainfall recording, the rain gauge was often sited on a roof or a wall, but our early rainfall observers soon discovered that rooftop gauges caught less rain – as much a 30 per cent less – than gauges at ground level, due to the effects of air turbulence. The modern rain gauge is inserted into the ground with its rim about 30 centimetres (12 inches) above the surface, but it is accepted that even these instruments catch up to five per cent less than the real amount of rain that falls. Rain is still measured in inches or millimetres over much of the world, although Spanish and Portuguese speaking countries use the more logical litres per square metre. The beauty of the metric system means that one millimetre of rain is also one litre per square metre.

During the late nineteenth century mechanical rain gauges were invented which allowed a continuous record of rainfall to be obtained over a specified period, usually a day or a week. The rain gauge funnel disgorges the rain into a container with a float which rises as the water level in the container rises. This float is connected to a pen which draws a line on a chart. The chart is attached to a clock-driven drum which rotates once per day or per week. From the line on the chart rainfall duration and rainfall intensity can be measured.

A place in the sun

Sunshine recording was a particular obsession of the British, probably because of the unreliability of that particular weather element in this cloudy corner of western Europe. Nowadays we have instruments called radiometers which measure the intensity of the light radiating from the sun, but for over a hundred years the usual means of monitoring sunshine was to measure the time that the sun shone each day, or even each hour.

Until the 1990s, the official worldwide standard instrument for this purpose was the Campbell-Stokes sunshine recorder. The original recorder was invented by the Scottish amateur scientist John F. Campbell in 1853, and it was refined by the famous physicist Sir George Stokes in 1879. It works on the same principle as the schoolboy who uses a magnifying glass to make scorch marks in his desk top by focusing the rays of the sun into a point. Instead of the magnifying glass it utilizes a solid glass sphere, with a piece of card in place of the desk top. The card is chemically treated so that it does not ignite, and the sun's heat burns a narrow trace along the length of the card, which has time marks every half-hour. Campbell's prototype used a spherical glass bulb filled with water, supported within a hemispherical wooden bowl, and the sun's rays scorched the wood. The scorch marks could be measured each day before being polished out. Daily measurements of sunshine were first made on an experimental basis from 1876 onwards at the two London observatories in Greenwich and Kew. But it was the new Campbell-Stokes instrument which made measurement simple and reliable enough to be entrusted to non-experts. Systematic sunshine recording began at several stations around the British Isles in 1880 or 1881.

When the wind blows

The first real meteorological measurements told us the direction from which the wind blew. Wind vanes have been popular for many centuries, and weather-cocks have adorned churches throughout Christendom for over a thousand years. The ancient Greeks thought of wind as the breath of the gods, and they named the different wind directions after some of their deities. The most important were Boreas, the north wind, Zephyros, the west wind, and Notos, the south wind. The wind vane forms the basic element of wind direction measurement to this day.

Wind speed has always been a much more difficult element to quantify. Very rough approximations have been possible by describing the behaviour of trees, soil particles, and smoke, while mariners closely studied the height and amplitude of the waves generated by the wind, as well as the deterioration in visibility caused by blowing spray. Early instruments tried to ascertain the force exerted by the wind by fixing a broad metal flap along one edge, placing it in an exposed location, and allowing it to swing in the breeze – rather like one of those mid-twentieth-century shop signs, or in Britain the ubiquitous pub signs. Ideally, this 'pressure-plate anemometer' should be attached to a wind vane so that it is always fully exposed to the wind, and hinged loosely along its top edge to minimize friction. The angle through which the plate is displaced by the wind is proportional to the force of the wind which in turn is roughly proportional to the square of the wind speed.

In 1846 the English instrument maker, John Robinson, invented a very different anemometer involving the now-familiar arrangement of hemispherical cups attached to a vertical shaft which rotates as the cups rotate. This forms the basis of many present-day anemometers. The number of times the cup-arrangement performs one revolution in a specified period of time is related in a rather complicated way to the speed of the wind, and this relationship can be used to calculate both the speed and the force of the wind for any period from three seconds to a day or more. A three-second wind is the official definition of a gust.

The family of observers

Unmanned electronic weather stations are beginning to supplant the human observer; but these stations are expensive, they need regular maintenance and calibration, and they are limited in what they can monitor. Temperature and pressure are straightforward, but an unmanned station has great difficulty measuring the amount of cloud, the type of cloud, and the difference between rain, sleet and wet snow. At the beginning of the new century, therefore, these electronic stations are economically viable only in rich countries and in very remote regions such as deserts, ice-caps and oceans. So there are still thousands and thousands of individual weather stations scattered across the world, staffed by weather observers who perform their duties reliably and regularly. It is important that they are both reliable and punctual because much of the value of their work lies in the worldwide collation of meteorological measurements all made to the same standard and all made at the same time. Every six hours these observations are made over the entire planet, providing a synopsis of the global weather at a particular moment – hence the term 'synoptic chart'. In Europe and in North America readings are made every hour, allowing hourly synoptic charts to be constructed for these continents.

Those observing posts which perform hourly (and sometimes half-hourly) observations are called 'synoptic stations', and are usually found at commercial airports and military airfields. This high frequency of meteorological reporting is needed because aircraft pilots and air-traffic control staff must know precisely in what sort of environment aircraft are taking off and landing. Clearly, wind and rain and visibility are vitally important, but so is air pressure, because an aircraft's altimeter is actually a special kind of barometer, and its zero setting is the barometric pressure observed at ground-level. Thus when the air pressure is changing rapidly, for example during the advance and retreat of a deep depression, it is vital that each pilot adjust his altimeter accordingly.

At the other end of the observing spectrum are 'climatological stations' where a single daily observation is carried out, usually at an hour around breakfast time. The job of these climatological observers is essentially to build up an archive of the climate at their particular location by making their one daily reading: logging the maximum and minimum temperature of the day, measuring the rainfall and sunshine and the run of the wind, and keeping a diary noting the salient features of each day's weather. Most synoptic stations perform the once-daily climatological readings alongside the rest of their functions. Evidently, climatological stations are not manned

24 hours a day; indeed, they are often visited once daily and the observation may be an additional 15-minute task for some functionary who has another job entirely. Such stations are found in schools and colleges, at coastal resorts and winter sports centres, at police stations and post offices, attached to coastguard stations and lighthouses, at sewage works and reservoirs, on farms and in managed woodland. And of course in many countries there is a band of weather enthusiasts who also keep their own weather observations, many of which are accepted as official weather records.

In some countries, notably in western Europe, these weather stations have been increasingly prone to vandalism in recent years, due to their open location and the lack of supervision. This, together with advances in technology, is encouraging their replacement by heavy-duty, vandal-proof electronic weather stations. Nevertheless, the familiar Stevenson screen – likened by some to a beehive on stilts – with thermometers inside and rain gauge nearby is likely to remain in regular use around the world for several decades to come.

Measuring the upper air

Nineteenth-century meteorologists learned a lot about weather patterns near the Earth's surface, but, with a few honourable exceptions, they knew little about what was happening way above their heads – in the upper reaches of the atmosphere.

There were, though, a few remarkable characters who understood that the upper air held the key to a much more complete understanding of the way the weather worked, and that through better knowledge would come better forecasts. Some established observatories on mountain tops, such as Mount Washington in the USA, Zugspitze in Germany, and Mont Aiguoïlle in France. In Paris, weather-recording equipment was fixed to the top of the Eiffel Tower upon its completion in 1889, while other scientists experimented with instruments attached to tethered balloons. The Mount

Washington observatory in New Hampshire, which it is claimed has 'the worst weather in the world', was first established in 1870, but the buildings were demolished by high winds seven years later. It was not rebuilt until 1932, but regular observations have been carried out ever since.

In Britain a mountain-top observatory was established in the 1880s on the summit of Ben Nevis, the country's highest mountain, under the auspices of the Scottish Meteorological Society. But several influential scientists in the UK were unconvinced that these summit observatories would actually provide data which were representative of the free atmosphere, and the Ben Nevis team struggled to finance their project, finally losing the battle in 1904. Long before that observatory was built, the British Association sponsored a series of manned hot-air balloon flights, with a weather observer making meteorological readings at regular intervals throughout the voyage in order to learn more about the temperature, humidity and pressure at various levels above the ground. They were particularly interested in measuring these elements in and adjacent to cloud layers. These balloon flights took place between 1862 and 1866, and an additional series of observations was made from a tethered balloon at Chelsea in London during 1869.

The man who undertook most of the observational work during these flights was James Glaisher, who was Superintendent of the Meteorological Department at the Greenwich Royal Observatory in London. The project almost came to a grisly end when one of the earliest ascents went too high, over 11 kilometres (7 miles) above the ground, and the occupants of the basket lost consciousness and almost died. Glaisher was undaunted by his experience, and carried on to complete the series of upper-air investigations. From the 1890s, sporadic upper-air soundings were made by attaching a mechanical apparatus known as a meteorograph to a tethered balloon. The meteorograph recorded temperature, pressure and humidity onto a rotating drum or disc which worked by clockwork. These instruments were prone to failure, they could not really respond quickly enough to the changes they encountered, the records were often damaged by moisture or ice, and they still could not record wind speed.

The radiosonde, a British invention, first took the skies in 1937, although it was after the Second World War before it really came into its own. Comprising a series of transducers to monitor the elements, a converter which transforms the readings into audio output, a switching mechanism so that each element can be measured in sequence, and a radio transmitter to send the data back to the ground receiving station, the whole package was attached to a helium-filled balloon. Tracking the balloon by radar allowed wind speed and direction to be calculated. Some of these sondes reached a height of 32 kilometres (20 miles) or more. By the late 1940s many countries around the world had their own radiosonde network, and to this day the global network still provides a substantial proportion of the upper-air data used in modern weather forecasts.

Chapter 13: Weather forecasting: from guesswork to super-computers

The Egyptian and Babylonian civilizations made concerted attempts to predict changes in the weather using a combination of sky-watching and astrology. The ancient Greeks brought rational thought to bear on the subject and Aristotle wrote a treatise entitled *Meteorologica* which demonstrated an understanding of the hydrological cycle and an appreciation of what clouds are. But from then until the seventeenth century meteorology, in common with many branches of science, remained in the sphere of natural philosophy rather than pure science. Aristotle's work, translated into Latin via Arabic, remained the only serious attempt at understanding the atmosphere for over 1800 years, and it was still quoted as an authority as recently as the beginning of the nineteenth century. Throughout this long period, weather forecasting was the preserve of countrymen, churchmen, and rogues, and it was during the Middle Ages in particular that the vast body of country weather lore accumulated.

As we have seen, scientific method was first of all brought to bear on meteorological observation, with instruments devised to measure all sorts of atmospheric characteristics from the late sixteenth century onwards. A hundred years later, in the 1680s, the mathematical groundwork for future meteorological study was laid by the German mathematician Gottfried Leibnitz who developed the calculus, and the English physicist Isaac Newton who proposed a new branch of mathematics involving the motion of objects called mechanics. Without these, the scientific study of the air in motion, first described by Leonhard Euler, a German mathematician, could never have begun. The American William Ferrel was the first to formulate equations to describe atmospheric motion on a rotating Earth in 1859.

Alongside these scientific developments there also came important technological advances. Vitally important was the invention of the electric telegraph by the American Samuel Morse, which enabled weather observations to be transmitted to a central office within minutes. Actual weather distribution could now be analysed in real time, and at last meteorologists were able to recognize patterns of weather which moved across the country from one day to the next, sometimes intensifying, sometimes dissipating. Further important steps occurred in 1850 with the linking of France and England by telegraph, and in 1866 with the first successful transatlantic telegraph cable. The German physicist Heinrich Brandes compiled the first synoptic weather chart in 1816, but the data he used were already 33 years old! Elias Loomis was the first to draw a real-time synoptic chart in 1843, and the first published weather maps were sold for a penny each at the Great Exhibition in London in 1851. Weather reports were printed in the *Daily News* in England in 1848, but the first weather maps to appear in a newspaper were found in the *Washington Evening Star* in the USA in 1856.

Now came the bureaucrats. In the 1850s the French and the British established embryonic national meteorological services; the Dutch soon followed, but it was another ten years or more before the American Weather Bureau was born.

The Meteorological Department of the Board of Trade was formed in the UK in 1854 following the wreck of British warships in a storm off the coast of the Crimea. Its first chief was Admiral Robert FitzRoy, who 20 years earlier had captained the *Beagle* which took Charles Darwin on his voyage to South America (see Chapter 9). Initially, FitzRoy's work revolved around supplying instruments to ships in return for regularly telegraphed weather observations, and developing a network of some 40 observing stations around the British Isles, mainly in coastal locations. Arguably his biggest contribution to meteorology was the storm warning service which he began in February 1861. Strictly speaking this was not a forecasting service, as he only issued warnings when a gale had already been reported at one of the coastal stations. Storm 'cones' were hoisted in prominent positions at ports and harbours around the coastline so that they could be seen by passing ships, and the orientation of the cone indicated whether the wind was blowing from the southerly or the northerly half of the compass.

Before long FitzRoy became more ambitious, publishing a series of rudimentary forecasting rules in his *Weather Book* in 1863, and providing weather forecasts for the whole country in *The Times* later the same year. However, meteorology was still a primitive science and FitzRoy's forecasts were far from infallible. The newspaper's editor, initially a keen supporter, lost faith as scepticism grew amongst his readers, and the forecasts ceased to appear early in 1865. A couple of months later FitzRoy took his own life, although personal difficulties rather than depression at the failure of his forecasts were probably behind his suicide. The storm warnings for mariners continued, and continue to this day in the form of the gale warnings and shipping forecast broadcast three times a day by the BBC.

In the USA the Weather Bureau was set up several years later, but it grew much faster than its British counterpart. Established in November 1870 as a subdivision of the Army Signal Service, it was running almost 300 outstations by the end of the 1870s and producing daily forecasts for the entire nation. By 1880 its director claimed a 74 per cent accuracy rate in its predictions of rain, snow and temperatures, although a close examination of his method indicates that he was very generous in what he considered to be a correct forecast. In 1890 the US Weather Bureau left the military sphere and became part of the Department of Agriculture.

During the latter years of the nineteenth century most advances in the science of meteorology emanated from individual scientists working in a university environment. One of the biggest advances came from the Norwegian mathematician, Vilhelm Bjerknes, whose circulation theorem published in 1898 at last showed an understanding of how air temperature, pressure, density and humidity could vary in three dimensions. Meanwhile the national weather services plodded away at the mundane tasks of collecting and analysing data, and producing not very good weather forecasts based on fairly crude rules of thumb.

One outstanding exception was the Director of the UK Meteorological Office from 1907 to 1920, Sir Napier Shaw, who was a talented experimental scientist, having been an Assistant Director of the Cavendish Laboratory from 1877 to 1906; after retiring from the Met Office he returned to academia to become the first professor of meteorology at the Royal College of Science (later Imperial College). He is credited with introducing a number of scientific skills, including the examination of temperature change with height, and also air-trajectory analysis, to the art of weather forecasting.

The coming of aviation early in the twentieth century delivered a huge impetus to the development of meteorology (see Chapter 9). The early aviators – the flying aces of the Great War and the civilian pilots in the years that followed – were voracious consumers of weather forecasts; it is difficult to imagine a more weather-sensitive occupation. Not only did they need a reliable picture of the wind and cloud conditions during their flights, but they also needed predictions of the weather at their expected destinations, and at alternative destinations in case they were diverted. At the same time aeroplanes and airships were able to provide new information about atmospheric conditions aloft, transmitting the data back to meteorologists on the ground by radio. Unmanned balloon flights were followed in 1937 by the invention of the radiosonde which sent back details of pressure, temperature, humidity and wind by automatic radio transmission.

During the Great War meteorologists prepared forecasts not only for the first military fliers, but also for army commanders in northern France. The first on-site meteorologists worked virtually in isolation behind British lines, measuring wind patterns in order to warn the troops of the threat of poison gas releases from the German front line. The war saw a rapid expansion in the number of weather observing stations across Europe, and meteorology finally took its place as an interesting and worthwhile area of investigation in universities.

Drawing on his wartime work, Vilhelm Bjerknes together with his son Jakob and a colleague called Halvor Solberg presented a radical new way of looking at weather charts in 1921. Known as the 'polar front theory of cyclone development', it regarded mid-latitude cyclones (or depressions) as three-dimensional features formed by the interaction of adjacent masses of warm and cold air; the boundary between the contrasting air masses was christened 'the polar front'. Bjerknes used the word 'front' in recognition of the similarity of the line of opposition between the two different air masses and that between two armies. Although now much modified, Bjerknes' polar front theory survives as a basic framework for understanding weather systems in temperate latitudes to this day. Indeed, those television weather maps with highs and lows, warm fronts and cold fronts, and flowing isobars, owe everything to the two Bjerknes and the Bergen School of Meteorology.

Just one year later Lewis Fry Richardson, an English meteorologist, published a book called *Weather Prediction by Numerical Process* in which he showed how pressure changes over a small area could be predicted by applying a set of fundamental mathematical equations governing atmospheric motion to a pre-existing pressure pattern. This was the advent of numerical weather prediction (NWP), which has done so much to improve the quality of weather forecasts during the last half-century. Richardson, however, was well ahead of his time. He did not, of course, have the benefit of electronic computers to solve the millions of equations necessary to produce a forecast for 24 hours ahead. His vision was of a time in the distant future when there would be huge weather forecasting factories, each containing a massive army of 64,000 mathematicians – human computers – whose sole task would be to perform all those inter-related calculations quickly enough to keep ahead of the changing weather.

The UK Meteorological Office began its research into numerical weather prediction in 1950, but it was not until 1965 that their first computer-generated weather forecasts were issued. Today, highly sophisticated NWP models, carrying out millions of calculations per second on data acquired from a fine-mesh three-dimensional grid of atmospheric observations, are run on very powerful super-computers by national weather services all over the world.

For many years the computer-based forecasts were frequently adjusted by human meteorologists who could see where the computer output was simply not possible, or outside previous experience. By the mid- or late 1970s the computer forecasts were, more often than not, better than those prepared by flesh-and-blood forecasters, and by the 1980s the necessity to intervene in the forecasting process on a routine basis disappeared. Even now, though, human intervention still happens from time to time, usually to adjust the starting conditions which are not quite right because of an erroneous weather report or a gap in the observations. More and more, the job of the human weather forecaster is to interpret the computer output in terms of the requirements of the customers and the geographical idiosyncrasies of the local area: computer models are not yet good enough to predict exactly where fog will form, where showers will occur, and what time clouds will develop or dissipate. Even here, though, progress is being made.

'Here is the weather forecast...'

We are all familiar with the weather forecasts for the public on television and radio and in the newspapers. But these are only a small proportion of the forecasts which are compiled and disseminated – and often sold – from the hundreds of meteorological offices around the world. Weather forecasts fall into four broad categories: nowcasts, short range forecasts, medium range forecasts, and long range forecasts. And the customer base can be also divided into four groups: worldwide public services, national public services, core services, and commercial services.

Nowcasts

Strictly speaking, these are very short range weather forecasts, covering the period from the present moment up to roughly six hours ahead. The available data used in the construction of these forecasts consist of the latest weather observations, satellite images, and rainfall radar displays – in other words, what the weather is doing *now*, hence the use of the term *nowcasts*. The fundamental premise of nowcasting is the extrapolation of present weather distribution over the next few hours, so that, for example, areas of rain are predicted to continue moving at the same speed as during the last few hours. Often these forecasting tasks are done by hand, although some forecast offices have now computerized the movement of rain areas on radar displays and cloud areas on satellite images for a few hours into the future.

Extrapolation, however, is a poor technique to use when trying to predict the time of clearance of fog, for instance, or the transition from rain to snow, and on these occasions there is still nothing to match the knowledge of an experienced human meteorologist. Simple extrapolation also has to be modified when the weather pattern is developing rather than just moving; for example, if the NWP model output indicates that a rain area is getting bigger and the falling rain is becoming heavier, then these changes also have to be taken into account.

The main consumers of nowcasts in the USA include local government authorities responsible for disseminating tornado, hailstorm, and flash flood warnings, and also radio and television stations which provide the same warnings for their public. Similar services are also prominent in Australia, South Africa, and Canada, but in the UK such violent and life-threatening near-instantaneous weather events are almost unknown.

Flash floods may be very rare in Britain, but serious flooding can develop within a matter of hours in the typically small river catchments of the country, and the fact that large areas are urbanized contributes to the speed at which rainwater finds its way into streams and rivers. The UK Environment Agency therefore uses detailed rainfall nowcasts which are fed into computer models of the various river catchments in order to predict the areas threatened by floodwaters, and the time when that threat is at its highest.

Nowcasts are also widely used at civil airports and military airfields to predict the formation and dispersal of fog and low cloud, and the organizers of major sporting events also benefit from the knowledge of very short-term changes in the weather. Every June in London this is seen to good effect during the All England Tennis Championships at Wimbledon.

Short range forecasts

The introduction of NWP models run on computers in the late 1960s brought a significant improvement in the quality of forecasts for one, two, three and more days ahead. Further improvements since than have come in fits and starts, with sizeable advances usually taking place when refinements to the computer models are made. By the turn of the century the degree of error in forecasts for 24 hours ahead was less than half what it was in 1970. In 2000, 72-hour forecasts were more accurate than 24-hour forecasts had been in 1970.

Forecasts for up to 48 hours ahead, therefore, are now detailed and usually accurate thanks to the huge amount of investment in computer power and NWP model development which has taken place during the last half century. When they go seriously wrong it is a newsworthy event. British meteorologists are still taunted by an inadequate forecast in October 1987 when a destructive gale swept southern England, including London. The correct response to this sort of mockery is to point out that, if an event 15 years before has to be dredged up as criticism, then the quality of weather forecasts since then evidently has been very high.

Short range forecasts provide much quantitative information on temperature, humidity, wind speed and direction, cloud-cover, the timing and intensity of rainfall, barometric pressure, and at sea also the height and period of wind-waves and swell. But the distribution of localized weather phenomena such as passing showers and patchy fog still has to be described qualitatively. Many users – the general public for instance – only really require qualitative descriptions, and here the skill of the human forecaster comes into its own.

Users of these predictions include the broadcast and print media (and through them the general public), military and civil aviation, marine transport and the offshore oil industry, farmers and horticulturists, civil engineers, electricity generation and distribution, shops and supermarkets, and the travel and leisure industries. In other words, just about everybody.

Medium range forecasts

The improvement of short range predictions has trickled down to periods of two to six days. In 1970 these were often worthless; nowadays broad predictions of the weather five or six days ahead are more often that not pretty accurate. They do, of course, go wrong more frequently than do 24-hour forecasts, but the occasions when the computerized forecast is most likely to be off target are often apparent to the human forecaster. To make predictions so far ahead requires data from the lowest 20 kilometres (12 miles) or so of the atmosphere over the entire planet; after all, a depression forming over Florida may be slowly dying in the Russian Arctic seven days later, having wreaked havoc across western Europe during the interim.

Several national meteorological centres run these models for up to 15 days ahead, although the output beyond about six days is only infrequently made available to paying customers or to the general public. Consenting meteorologists in private sometimes refer to these extended predictions as 'fantasy land' because they are correct less than half the time. But useful information can be gleaned from this sort of computer output when it forms part of what is called an 'ensemble forecast'. What happens is that the computer model is run over and over again with very small changes being made in the 'start' conditions. On most occasions the forecast charts for the first five or six days look practically the same, but then they begin to diverge around seven or eight days, and this divergence accelerates beyond that. Sometimes the divergence is small and most members of the ensemble predict roughly the same weather pattern ten or even fifteen days ahead; this is telling us that on these occasions the atmosphere is predictable, so a very broad-brush outlook might usefully be made. On other days the ensemble members may diverge dramatically after five days, telling us the atmosphere has little or no predictability, and eliminating any hope of a useful forecast beyond that point. On yet other occasions the individual forecasts of the ensemble may divide into two groups implying that the weather could follow either of two courses; the human forecaster then knows to look for evidence of which route the real weather pattern will take before he decides which forecast to issue.

The media are almost as interested in medium range forecasts as they are in short range ones. In some countries there is serious competition between the TV networks to provide not only the most accurate forecasts, and the longest-range forecasts, but also to be first to 'call' major weather events. Beyond a certain point such rivalry becomes counter-productive, ten-day forecasts of a blizzard or heatwave fail, and the general public is left with an impression of television meteorologists who cry wolf. In a completely different context we saw exactly the same sort of rivalry during the US presidential election in November 2000. Some of the networks ended up with considerable egg on their corporate faces.

Long range forecasts

We have seen that the usefulness of weather forecasts using NWP techniques dimishes rapidly after about six days, and the modelling experts tell us that there is almost certainly a limit to this sort of forecasting. They say that, even with computers hundreds of times more powerful than the ones we use today, vastly more intricate NWP models, and a much greater ability to measure what the atmosphere is doing at any given moment, we are unlikely ever to be able to predict the weather more than two weeks in advance.

Superficially, then, the idea that it is possible to devise any sort of useful prediction for longer periods might seem incompatible. However, there are certain features of the weather machine which are predictable on a longer time scale. The most obvious of these are the seasons – anyone can predict that July in London will be warmer than the preceding January, or that in Bombay August will be wetter than April. At a much less obvious level, these seasonal cycles cause changes in snow and ice cover, and in the distribution of sea-surface temperatures across the oceans, and these in turn often encourage certain types of weather pattern to occur more frequently than others.

Predictions for one month or three months ahead are routinely made by the national meteorological centres and private forecasting companies in the USA, the UK and some other countries. They are generally very broad-brush, indicating for instance how far the monthly temperature and rainfall will deviate from the long-term average; they do not attempt to predict the weather on individual days. Even so, the level of accuracy in these long range forecasts is much lower than is achieved on a shorter time scale.

Long range forecasting also attracts a number of eccentrics, some of whom are convinced that they have the one true method, notwithstanding repeated failure. Many are harmless, but there are a few frauds whose main interest is to acquire money from gullible customers. None the less, it is not beyond the bounds of possibility that one of these mavericks, working alone, could yet discover new scientific relationships which improve the quality and status of present-day long range forecasting techniques.

Chapter 14: Commercializing the weather

There was a world of difference in the way meteorological services developed during the last century between the USA on the one hand, and most other countries in the world on the other. National meteorological institutions were established in most nations during the second half of the nineteenth century and the first half of the twentieth, and the new nations which emerged from the decolonization process after 1945 were quick to follow suit. These institutions were universally staffed by civil servants and funded by central government, although there was some variation in the government departments in which meteorologists found themselves. Many were allied to the military: for instance, the UK Meteorological Office was for a long time part of the Air Ministry, and then from 1965 the Ministry of Defence. The National Weather Service of the USA finds itself in the National Oceanic and Atmospheric Administration (NOAA) of the Department of Commerce, while Météo France is an integral part of the French Ministry of Transport.

Only in America...

The crucial factor which determined a different course of development in the USA compared with elsewhere is the fact that American government employees are constitutionally barred from engaging in commercial activity. Thus the National Weather Service (NWS) of the USA has remained essentially a core service throughout its history, collecting and analysing observations, preparing specialist forecasts for other government departments and agencies, especially for aviation, and general forecasts for the public at large, and issuing warnings of life-threatening meteorological phenomena such as hurricanes and tornadoes. All data was, and still is, available at minimal cost to academics, commercial meteorologists and the general public.

Today the NWS's mission statement is as follows:

> To provide weather and flood warnings, public forecasts and advisories for all of the United States, its territories, adjacent waters and ocean areas, primarily for the protection of life and property. NWS data and products are provided to private meteorologists for the provision of all specialized services.

In order to achieve this, the NWS is required to:

1. Co-ordinate programmes with state, local and federal agencies involved with meteorology and hydrology to attain maximum cost effectiveness. For example, the NWS works with aviation safety and forest fire prevention schemes.
2. Provide a spectrum of weather services to the private-sector meteorological and hydrological communities.
3. Provide data and forecast products to the private sector.
4. Work closely with the mass media as the chief means of communicating weather and flood warnings and forecasts to the public.
5. Fulfil international hydrometeorological obligations.
6. Conduct applied research with other agencies and the scientific community to improve warnings and forecasts based upon scientific and technological advances.
7. Enhance dissemination and information exchange services.
8. Facilitate improvements in the emergency management decision process.

The NWS was completely modernized and restructured during the 1990s, following a period in which it had fallen alarmingly far behind other national meteorological offices in the incorporation of computer technology into its services. Computers and satellites were upgraded, new communications systems were introduced, electronic weather monitoring stations were installed at over 300 sites across the country, and a network of advanced radars for tracking heavy rain, hailstorms and tornadoes was established.

The non-commercial nature of the NWS has resulted in the growth of a thriving meteorological private sector in the USA. A recent edition of the *Bulletin of the American Meteorological Society* contained advertisements for 128 companies and private consultancies engaged in meteorological work. Many of them concentrate on a single specialism such as meteorological instrumentation, pollution monitoring, and forensic meteorology (that is, the provision of evidence and expert opinion in courts of law). Others offer a wide range of services including forecasts for aviation and marine operations, weather modification projects, provision of specialist weather data, evaluation of weather at accident and disaster sites, and so on. Those 128 advertisers are only the tip of the iceberg: many other meteorologists work in larger organizations such as oil companies, civil engineering companies, radio and television stations, newspapers, legal firms, stockbrokers, in crop-spraying and hail suppression programmes, and in forestry and agriculture.

The American Meteorological Society operates a certification programme which is designed to establish and maintain a high level of professional competence, mature judgment, and ethical commercial behaviour. The successful applicants have to prove, by examination and interview, a thorough knowledge of meteorological science, considerable practical experience, and good character. The scheme has, broadly speaking, achieved its aim of discouraging the incompetent, the deceitful, and the irresponsible.

One of the results of the NWS modernization programme of the 1990s was a sharp reduction in the number of government employees in this sector. This followed the replacement of manned observing stations with automated ones. It is very probable that there will be further rationalization and modernization schemes during the early decades of the new century, which should make further savings on staff costs. The number of forecasters is expected to drop progressively as more and more services are automated, and more and more customers (such as military and civilian pilots) turn to self-briefing. Centralization of some services will also contribute to this process. Some of the loss in staff numbers will be absorbed by retirement, but many former government meteorologists can be expected to move sideways into the private sector.

The meteorological monopolies of Europe

In the larger European countries the ambit and activity of the national meteorological institutions are very different from those of the NWS in America. Until the 1960s there was really very little difference; the meteorological offices of France, the United Kingdom and Germany performed the same core functions as the NWS did, observing the weather, gathering data, producing forecasts and warnings, and providing specialist information to a handful of customers at low cost. There were very few commercial outlets for meteorological data and forecasts, and there was no appreciation in these organizations that there was money to be made, notwithstanding the burgeoning private sector across the Atlantic. As far as they were concerned, that was America, and therefore irrelevant to their circumstances. More to the point, there was no pressure on these meteorological offices from the respective central governments to generate income, and they continued to be funded from the public purse with little question.

Three things happened between 1965 and 1975 which changed everything. Oil and natural gas were discovered in the North Sea, and to a lesser extent in the Mediterranean basin; the advent of numerical weather prediction led to a demand for appreciably more money to fund technological advances in meteorology; and the Middle East war and subsequent oil crisis of 1973–74 led to recession and a rigorous cost-cutting régime in all departments of government. One way to appease their respective political masters was to start charging more for the services they provided.

Before that happened, however, first natural gas and then oil were discovered in the sedimentary rocks beneath the North Sea. Oil companies established offices in the UK, the Netherlands, Denmark and Norway, but it was in the UK sector that exploration really began in a big way during the mid-1960s. Aberdeen became the oil capital of Europe from the early 1970s onwards. Oil exploration in the North Sea was very different from what had happened before in marine environments. In the Gulf of Mexico, offshore Venezuela and in the Arabian Gulf rigs drilled in shallow waters during long periods of benign weather. (They avoided the hurricane season in the Gulf of Mexico, of course). In the North Sea natural gas was discovered in shallow waters off the coasts of East Anglia, Lincolnshire, and the Netherlands, but the search for oil was concentrated north of the 55th parallel in relatively deep water where storms were frequent and and calm seas practically unknown. By the mid-1970s rigs were operating in water depths approaching 300 metres (1000 feet) in the West Shetland Basin – that is, in the most storm-battered region of the Atlantic Ocean. A widely-quoted saying of the time uttered by a hard-bitten American oilman newly arrived in Aberdeen was: 'They told me you don't have hurricanes here; well you darn-well do, but you jus' call 'em depressions.'

Every winter these depressions pass through the North Sea, bringing prolonged gales with sustained wind speeds of up to 70 mph and peak gusts in excess of 100 mph. Wave heights of 10 metres (33 feet) or more are common in this season, waves 15 metres (50 feet) high are occasionally recorded, and one of 29 metres (96 feet) was measured during a storm in 1976. Thus drilling rigs and drill-ships had to work in some of the most inhospitable waters in the world, and accurate weather forecasts were essential to the safe and efficient operation of these vessels. Many of the oil-companies naturally took their requirements to the UK Meteorological Office, but the service obtained from them was less than adequate. The first forecasts lacked detail, were prepared by aviation forecasters many of whom had no marine experience at all, and the detailed climatology of the open waters of the North Sea (in marked contrast to the lands surrounding it) was practically unknown. Worse, the Meteorological Office provided their service from the London Weather Centre.

Private-sector companies had already been forecasting for the oil industry in the Arabian Gulf since the late 1950s, and as this region was then very much in the British sphere of influence, they were mainly British companies staffed with British meteorologists. They quickly established small offices in East Anglia and Lincolnshire, running on a shoestring, when rigs started working in the southern North Sea, and it was not long before they moved to Aberdeen as the focus of offshore activity transferred to the central and northern parts of the North Sea. Thus they provided immediate competition with the Meteorological Office service. For many years it was no contest: the private sector companies were at the commercial centre of the North Sea oil boom, they had forecasters with marine experience including many former Royal Navy meteorologists, and they frequently provided forecasters to work at the sharp end – actually on the rigs, briefing the oilmen on site, face to face. Being so close to the action they also learnt quickly about particular features of the North Sea climate, and they presented much more detailed information on wind speed, wave height and period, and swell characteristics, than their government-based competitors could. These, then, were the first successful private sector weather forecasting organizations on the European side of the Atlantic.

The rest of the North Sea story is a sad one. The UK Meteorological Office caught up with the private sector forecasting companies by about 1980, opening an Aberdeen office, and pumping millions of pounds of research into improving their meteorological knowledge of the North Sea and the behaviour of wind-waves and swell-waves there. There then followed a price-cutting war which eliminated most of the private sector competition because the state meteorological service was able to subsidize its North Sea operation from public funds. By 1986 North Sea weather forecasts were much more detailed and much more accurate than they had been in 1973, but they cost the oil companies £30 per day compared with £50 per day 13 years earlier. Taking into account inflation – the Retail Price Index shows that prices quadrupled during that period – these forecasts were sold in 1986 for less than one-sixth of their 1973 cost.

Nevertheless, the die had been cast. It was in the UK and the Netherlands that weather forecasting companies and consultancies were most successful during the 1980s and 1990s, making early inroads into the insurance and media markets. By the end of the 1990s they were offering a wide range of services to all sectors of the business community, although even in Britain they still accounted for less than five per cent of the turnover. In some countries, notably France, Germany and Italy, the private sector faced particularly hostile state meteorological services, and progress here was even slower.

In the UK the extreme hostility demonstrated towards the private sector by the Meteorological Office during the 1970s and 1980s mellowed somewhat during the last decade of the twentieth century. Ironically, private sector meteorologists had opened up a number of sectors which the Met Office had previously been providing a minimal service for, or which they had ignored altogether. Notwithstanding an improving relationship between the two sectors, there were still deep disagreements. The most damaging of these concerned the provision of basic data. Already paid for by the taxpayer as part of the core service of the various European state meteorological services, these national institutions initially refused to make much of the data available, and subsequently offered to provide it at an outrageous cost. In 1993 the majority of these national offices joined forces to launch ECOMET, a grouping designed to co-ordinate the sale of meteorological data to private weather forecasting companies, although they protested loudly – perhaps too loudly – when accused of operating a price-fixing cartel. ECOMET established a secretariat in Brussels, and it is doubtful that the income derived from selling data has covered even the cost of this secretariat in any year since its creation. This was an untenable situation as far as the private sector was concerned, and they came together to found the AIWS, the Association of Independent Weather Services, in order to fight what they regarded as the outrageous prices sought by ECOMET.

By the early years of the twenty-first century the position of the state services was beginning to erode. Several European weather forecasting companies established a presence in the USA, while others merged with or were taken over by American corporations, and were therefore able to benefit from the free access to data there. Because all data are freely available to all national meteorological offices, the NWS in America has access to all European weather observations. This led to the ludicrous situation of, for instance, a British weather forecasting company based in Birmingham setting up an office in, say, San Francisco in order to acquire the weather observations made at Birmingham airport. The leakage of all manner of data onto the internet has helped to accelerate that erosion, and it now seems likely that most meteorological data will become available at minimal cost pretty well worldwide during the next decade or so.

Chapter 15: Towards a free market in the UK

The UK Meteorological Office was directly administered and funded through the Ministry of Defence until 1990, but under the free-market economic policies of Margaret Thatcher's Conservative government it joined several other departments of state in having the strings tying it to government progressively loosened. Until the mid-1990s it still benefited from direct funding from the Ministy of Defence, to the tune of £83.4 million in the fiscal year 1995–96. In 1996 it became a 'trading fund', enabling it to operate its commercial services almost as a private company might do, and the defence ministry became one of its customers.

There is still desultory talk – both in the meteorological community and amongst politicians – of outright privatization, while retaining the core service as part of central government. The present trend indicates that privatization may eventually happen almost by default as the Office's commercial operations become more and more autonomous. On the other hand, serious and protracted discussions have been taking place during 2001–02 on a tripartate basis between ECOMET, the Association of Environmental Data Users or AEDU (a successor to the AIWS), and the European Commission's competition secretariat; the AEDU argues that the only way to provide a level playing field for competition between the national meteorological offices and private weather companies is to completely privatize the commercial sectors of all of the national offices. The ECOMET representatives are fighting very hard to prevent this happening, but at the time of writing the European Commission seems inclined to be persuaded by AEDU's arguments. However, there is still a big gap between the Commission being persuaded and the Commission instructing European Union member governments to proceed with privatization. There is an even bigger gap between individual governments being told to do something and actually doing it.

In 1989 the Social Affairs Unit, a right-wing think tank, published a paper proposing the gradual privatization of weather forecasting in the UK, written by Dr Jerome Ellig, an American economics professor who had been a consultant on privatization to the presidency during Ronald Reagan's term of office.

Ellig pointed out that the debate on privatization in Europe was strongly tilted from the outset by the widespread belief in most European countries that 'only governments can gather data and prepare reliable weather forecasts'. He added that there is little economic justification for retaining these activities under the government's roof. Private weather forecasting companies live or die by the accuracy of their product, as well as by the quality of all aspects of the service they provide. A monolithic government institution, especially one with a virtual monopoly, is rarely responsive to the needs of the customer; indeed, government involvement in the weather business 'deprives citizens of receiving full value for their money because it prevents the evolution of more effective weather services'. He also criticized the Meteorological Office for its pricing policy, especially in respect of the data costs to private companies where the Met Office acted as a monopolistic manufacturer, wholesaler and retailer rolled into one.

From the standpoint of 1989, Professor Ellig admitted that, however strong the economic arguments for privatization might be, 'they are doomed to defeat without a politically realistic privatization plan ... not even the Thatcher government, long a champion of privatization, has been willing to propose this alternative for the Met Office.' One problem, he said, was that there was relatively little experience of private weather services in the UK, so a successful privatization would have to proceed gradually. He foresaw a core service, rather like the American National Weather Service, remaining under the direct control of government.

It is amusing to read the UK Meteorological Office's response to the Social Affairs Unit report after almost a decade and a half of grudging and hesitant progress down the privatization path. At once both arrogantly dismissive and deeply aggrieved that anyone could believe it was not for the best in this best of all possible worlds, Peter Ryder failed to address any of the economic arguments that Professor Ellig had advanced. There was no appreciation at all – or none admitted – of the illogicality of reserving data paid for the taxpayer for its own commercial purposes and effectively denying it to the embryonic private sector of the time. Instead he claimed that 'the Met Office does not run a monopoly; there are quite a number of private weather companies in this country already. We have had experience over many years of providing commercial services in fair competition with that private sector. We believe that the UK Met Office, which combines and integrates all the aspects of meteorology from basic research to commercial services, is both efficient and effective.' Within nine months the first step towards the real world had taken place when the government transformed the Meteorological Office into an Executive Agency.

By 2001 the language had changed. In his annual report, the Chief Executive wrote: 'Meeting our customers' needs in a cost-effective and timely manner remains at the heart of our business strategy. Expectations regarding price and quality continue to grow . . . we are not only committed to meeting this challenge but also to looking at ways to improve our responsiveness as the requirements of our customers become more demanding.' Annual Report-ese, certainly, but written with the experience of several years of quasi-commercial activity under its corporate belt. And yet there are still plenty of reminders of the old attitude. The Meteorological Office has been extremely reluctant to make available to the general public its rainfall radar information, an excellent product largely funded by the taxpayer, and they kicked up an enormous fuss when the BBC, which purchases the information for broadcast on television, also published it on its website. Having lost that argument the Met Office quietly followed suit within months, and real-time radar images are now to be found on their website as well.

The men and women who run the companies and consultancies in the private sector have appreciated the change in attitude, at least at a superficial level. Instead of being cold-shouldered and branded 'cowboys', they are now invited to presentations, occasionally wined and dined, and generally treated as human beings. But beneath the surface the unspoken belief remains that the Met Office should be the only provider of meteorological services to commerce and industry. A certain level of arrogance can be found on both sides, however. The chief of a private sector forecasting company recently said that, given the hugely unfair advantage it has built for itself, the Met Office really ought to have seen off the competition years ago, and the fact that it has not merely shows that they haven't a clue about the commercial world: 'they are like children playing shop'.

In 1989 the Met Office's Dr Ryder made great play of the fact that he and his colleagues believed they were operating 'in fair competition' with the private sector. The annual financial statement for the fiscal year 1986–87 – not exactly the most transparent set of accounts in the world – indicated receipts of £23.6 million and 'customer activity costs' of £21.1 million, but at that time those costs did not appear to include the expenditure involved in gathering data. These were the self-same data for which the Met Office were attempting to charge private weather-forecasting companies a 'commercial price'.

The accounts and financial information for the fiscal year 1999–2000 were rather more detailed, and revealed some interesting facts. Turnover had by then increased to £151 million, costs (including operating expenses) stood at £142.6 million, resulting in an operating profit of £8.4 million. Nearly half of the turnover, £71.3 million, was accounted for by what are called services to 'core' customers. The 'core' is defined as 'the programme of work necessary to generate, and make available centrally, the underpinning weather forecasts and climatological services which are the basis for specified direct services to core customers'; the core customers are primarily the Ministry of Defence (37 per cent), other government departments (37 per cent), and civil aviation (24 per cent). This 'core', of course, also provides the underpinning weather forecasts and climatological services which are the basis for services to all other customers.

The remaining turnover, amounting to £79.7 million, is accounted for by what are described as 'direct services' to a wide variety of customers, and these benefit from all of the high-quality data and forecasts already paid for by the core customers. However, little more than one-quarter of these 'direct services' is described as commercial and therefore, in the Met Office's own words, 'subject to open competition'.

There we have it, then. After more than a decade of what it considers to be 'fair competition with the private sector', a mere 13.6 per cent of the Met Office's income is from contracts open to competition from private firms, increasing to 26.1 per cent when one excludes so-called 'core' contracts. The core activity of collecting data and creating forecasts is paid for by 'core' customers, it is available to that sector of the Met Office providing commercial services in open competition with the private sector, but most of it is not available to those private meteorological companies.

The financial summary of the commercial sector of the Met Office reported a turnover of £20.9 million in 1999–2000 of which £0.3 million – just 1.5 per cent – is rather coyly described as a 'contribution to the core'. This contribution is equal to the operating profit of the commercial sector, and presumably would be zero if the account were exactly in balance. One wonders whether the core would contribute to commercial services if they reported a loss. The published statement of accounts unfortunately does not provide any detail of the turnover and expenditure in the commercial sector, so it is impossible to determine what sort of careful accounting renders the operating profit close to zero year after year. However, any business or consultant operating in the private sector would be delighted to be offered full access to the end-products of the Met Office's core activity for 1.5 per cent of their turnover.

From this analysis, two fundamental questions arise for those who still think that the Met Office is involved in fair competition with the private sector:

1. Why are 74 per cent of the Met Office's non-core services not available to open competition?
2. Why are private companies not given the same access to the products of the Met Office's core activity as its own commercial sector is, and at the same cost?

And we might ask a third, rhetorical, question: given this sort of competition, is it any surprise that, almost 40 years after the establishment of the first private forecasting company to service oil companies operating in the North Sea, the private sector still accounts for less than five per cent of meteorological services provided in the UK?

SECTION THREE
A change in the weather

Chapter 16: Fixing the weather

We have all had moments when we would have liked to switch on the sunshine or switch off the rain, to slake a drought, or to adjust the outdoor thermostat. Ever since early man set up home in the nearest cave, or lit the first fire, we have endeavoured to adjust our personal microclimates – you could call these the first attempts at weather control. Today, weather control might be thought of as the preserve of mad scientists and second-rate novelists, but there are some kinds of weather modification, albeit limited in scope, which are routine in certain areas of human activity.

We should here distinguish between deliberate and inadvertent modifications, and also between altering the weather and altering the climate. There are also differences of scale which will come into the discussion and which have an important bearing on whether such modification is socially and politically acceptable.

Deliberate attempts to modify the weather include, at the most local level, the dissipation of fog on airfields, and, at the most ambitious level, the dispersal of incipient hurricanes. Inadvertent changes range from the triggering by power station emissions of cumulus cloud development which eventually produces rain showers, to the prolonged, heavily polluted smoke fogs which used to plague European and American cities before the introduction of clean air legislation.

As for climate, we know that the emission of greenhouse gases is held responsible for the unpremeditated changes which are presently showing their hand. However, planned climate changes might once have included the dusting of ice and snow with coal dust to induce large-scale melting of polar ice-caps in order to forestall a global cooling.

Why change the weather?

There is one powerful reason to seek to alter the weather, and that is to reduce the losses, both human and economic, wrought by a natural hazard or disaster. There are other reasons but these do not have the same moral grounding: they may include a desire for economic or even military advantage over a neighbour. Saving human lives and property can be achieved either by modifying the hazard itself, or by reducing its impact. These two methods need not be mutually exclusive, and in certain cases a combination may provide the best results. Insurance may also help to soften the blow of damaged property or loss of earnings.

Meteorological disasters are generally less open to control than other catastrophic events such as landslips and avalanches where large-scale engineering projects can reduce the frequency and the intensity of the hazard. The destructive forces of the atmosphere, however, are massive in comparison with the degree of control which human beings are yet able to exert over them. To illustrate this we can make a rough estimate that a typical winter mid-latitude depression over the Atlantic or Pacific or Southern Ocean releases about 100,000 times as much energy as did the first atomic bomb. Even if we had the inclination to interfere with the atmosphere on this scale, the cost-benefit analysis would soon see the project terminated. There are, though, certain circumstances when intervention can work: when the atmosphere is finely balanced between two contrasting states, we can successfully tip that balance with relatively low cost and effort. Thus cloud-seeding operations to create rain will never work if the weather-machine is not pre-disposed towards the development of clouds throughout a considerable depth of the atmosphere, but they may succeed if the atmosphere is in a condition in which deep cloud development is already occurring.

Weather modification projects, whether successful or not, bring a variety of problems which make the prospect of large-scale activity in this field very uncertain. Most of the chemical agents used in seeding are themselves pollutants, and the disadvantages of these may outweigh the advantages achieved by the seeding process. The release of large amounts of heat energy to disperse fog or cloud, or the generation of extensive electric fields to disrupt storms may fall into a similar category. But most problematic of all is our inability to restrict the effect of the modification to the area for which it was intended. For instance, suppressing a hailstorm over a region of farmland may displace the storm activity to a neighbouring urban area; the crops are saved, but in the city windows are smashed, motor-cars damaged, and people seriously injured. One scarcely dares to imagine the legal repercussions of such an event.

Even more contentious would be weather modification which crossed international boundaries. For instance, dissipating a hurricane which was threatening, say, the Texas coastline could conceivably result in the development of a new hurricane which sweeps through the central Caribbean and makes landfall on the Mexican coast, causing billions of dollars of damage and costing hundreds of lives. What sort of international tension might follow this sort of catastrophe? Nor does it stop there. We know that hurricanes and typhoons are one of the principal mechanisms whereby heat energy received (from the sun) in the tropics is transferred to middle and high latitudes. If that mechanism were interfered with, who knows what sort of chain reaction might follow? What kind of major climatic anomalies might turn up in different parts of the world during subsequent months?

At our level of technology more worthwhile results – cheaper, too – are best achieved, not by attempting to control a destructive weather event, but by modifying the vulnerability of human communities. This includes a wide range of activity, such as building physical protection like dykes, sea-walls, levées and overflow channels; changing land-use to diminish for example the likelihood of flash floods, mudslides and avalanches; providing a legal framework to discourage or prevent urban development in at-risk zones such as flood-plains of rivers and low-lying coastlines, and offering financial inducements for people already living in these areas to move; raising building design standards so that homes and commercial premises can withstand violent gales or hurricanes; increasing the awareness of the community to the potential hazard and investing in forecasting and warning services; and making available adequate insurance facilities for those affected by a natural disaster. The cost-effectiveness of all these activities can be simply demonstrated, but even in the world's richer countries the political will to enact them is haphazard and half-hearted. Indeed, powerful vested interests, both political and commercial, often stand in the way of common sense action.

Weather wars

Altering the weather to wage war is a chilling idea, but we have seen in an earlier chapter how important the weather has been in past conflicts, so it should not surprise us that military leaders might seek to benefit even from some rudimentary control. Indeed it has already happened. Some of the protagonists during the Great War (1914 to 1918) released poison gas to disrupt the enemy's front line: this was in effect a crude alteration of the chemical composition of the lower atmosphere, and therefore amounted to a meteorological modification. Furthermore, both sides used meteorologists on site to observe and predict the wind patterns, in order to maximise the effectiveness of the poison gas release on the one hand, and to instigate precautionary measures on the other. During the Second World War (1939–1945) frequent attempts were made to dissipate artificially fog and low cloud at military airfields to increase the number of flying hours; conversely on clear, moonlit nights (especially with a fresh snow-cover which reflected moonlight and therefore improved visibility) artificial smoke fogs were created in urban areas and at other potential military targets in an attempt to confuse enemy raiders.

Chapter 17: Making rain and killing hail

Since the earliest farmers, some five thousand years ago, human beings have raged against the destructive powers of torrential rain and hail, and have sought ways of protecting their crops from ruination. Hailstorms in particular are merciless, and even in the twenty-first century the modern farmer can do precious little to prevent widespread damage and considerable financial loss. Insurance premiums in districts prone to severe hailstorms can be prohibitively high, and in some countries hail damage policies are simply not available – such a storm is considered to be an act of God.

Frightening the hail gods

The earliest communities tried shooting arrows into clouds to scare the spirits which lived there and which were responsible for making the hail. In central and southern Europe church bells were rung when storms approached, although the continuing belief that evil spirits inhabited storm clouds was probably heretical. European travellers to the Far East in the fourteenth and fifteenth centuries observed Chinese peasants firing rudimentary cannons into the clouds, an activity which probably began shortly after the discovery of gunpowder during the tenth century. The idea was taken up in Europe especially in Austria, Italy and France during the 1600s and 1700s. It was impossible to determine how successful the activity was, and it was around this time that the first legal disputes arose: vineyard owners accused each other of diverting hailstorms onto their land.

An Austrian farmer called Albert Stiger who became the mayor of his town established a network of three-centimetre guns throughout his valley. These guns pointed directly upwards and had broad stacks attached in order to amplify the noise of the explosions. When heavy cumulo-nimbus clouds developed the guns were fired systematically and repeatedly until the storm threat had passed. Stiger set up the network following several summers with frequent damaging hailstorms, yet the first year of operation – 1896 – was almost completely hail-free. This was probably coincidence, but it was understandable that Stiger's guns achieved widespread notoriety, not just in Europe but in Russia and the USA too. Within five years there were some 10,000 Stiger guns operating in over 20 countries round the world. Whether or not they worked was a matter of fierce debate at the time, but they were certainly dangerous devices, and many people were killed or injured when the shot returned to earth. The method was discredited by scientific testing during 1902, but even today hail cannons can be found in various parts of southern Europe.

More recently there have been several experimental programmes involving the firing of rockets containing cloud-seeding nuclei into storm clouds. This included a British trial in Kenya in 1963 just before that country's independence to try to reduce losses on tea plantations. Soviet experiments took place in the Trans-Caucasus during the early 1970s, and there were some in northern Italy around 1974–75 as well. The Soviets claimed an 80 per cent success rate, but post-Soviet examination of the research papers indicates that the figures were extensively massaged.

Today, small scale vineyards in Italy protect their vines by covering them with reed matting during thundery weather, while peach and nectarine orchards in south-western France use a fine-mesh plastic or carbon-fibre netting for the same purpose. Such protection is all very well for the relatively small farms which are still in a majority in many European and Middle Eastern countries, but it is clearly impossible to install where cultivation takes place on a very large scale, as in the USA, Canada and Russia.

Raindrops keep falling...

In areas of unreliable rainfall the research emphasis has been on artificially creating rain, rather than disrupting major storms. The first cloud-seeding experiments were conducted by American scientists in July 1946, and the heyday of these seeding programmes lasted from the late 1940s until the mid-1950s. The work focused on individual cumulus clouds, particularly those which were growing towards maturity and therefore had the potential to produce rain. It was accepted that the prospect of inducing rain to fall from small, fair-weather cumulus clouds was minimal, while the likelihood of being able to make any significant contribution to the rain-generation processes in large-scale cloud systems associated with depressions or fronts was also extremely small. Experiments continue to this day, although at a much lower level of activity compared with 50 years ago.

The general theory of how rain is generated was largely established in the 1930s and 1940s, and much is owed to the American physicist and chemist Irving Langmuir, the Swedish meteorologist Tor Bergeron, and the German physicist Walter Findeisen. Langmuir demonstrated that clouds contain water droplets of various sizes; as these drops fall under the influence of gravity the larger ones fall faster than the smaller ones, often colliding with them and absorbing them. Eventually the larger droplets 'scavenge' most of the smaller ones and become big enough and heavy enough to fall out of the cloud as raindrops. This process became known as the Langmuir chain-reaction theory, but is now generally called the 'coalescence process'.

Bergeron observed that most of the clouds which produced rain contained ice crystals, although these tiny particles of ice usually coexisted with water droplets. Both the ice and water particles form on what are called 'condensation nuclei' – microscopic fragments of dust or sea-salt which are found throughout the lower atmosphere. However, in a cloud where both ice and water are present, the ice crystals will grow faster than the water droplets because they are more efficient at absorbing water vapour within the cloud. They will also grow by colliding with each other, and by colliding with water droplets, eventually forming substantial snowflakes. The snowflakes may melt before they leave the cloud if the temperature at the base of the cloud is high enough and they may melt on their journey between the cloud and the ground, in both cases resulting in a shower of rain. If the air temperature is very low all the way down the ground a snow shower will be the outcome. Thus most rain that falls on us is melted snow, but snow is never frozen rain. This mechanism is normally called the 'Bergeron process' or the 'Bergeron-Findeisen process'.

For many years – until the 1980s – it was generally accepted that the Bergeron process was the agent of rain-making much more frequently than the coalescence process. Today, however, it is understood that both processes play vital parts in the creation of rain in shower clouds, thunder clouds, and extensive cloud-systems associated with fronts and depressions.

The theory behind the artificial stimulation of rain is simple: we can help a cloud to drop its load by supplying it with condensation nuclei which work more effectively than those which occur naturally in the cloud, or we can inject large water droplets into the lower part of a cloud which does not possess sufficient numbers of them to trigger the formation of raindrops. Let us briefly examine each of these methods.

Those American experiments which began in the late 1940s attempted to initiate ice crystal formation by sprinkling cumulus clouds whose tops have a temperature below zero Celsius (32°F) with pellets of dry ice (frozen carbon dioxide) or silver iodide or sodium chloride (salt) crystals. The theory shows that clouds treated in this way will produce rain at an earlier stage in their development than would be the case under normal circumstances. But the early seeding experimenters did not know how much seeding was necessary for this to work, nor did they know how many of the clouds which produced rain after seeding would have developed sufficiently to produce rain anyway. It was a Catch 22 situation. If there was no way of knowing whether their efforts were successful, they could not know whether or not enough of the seeding chemical was being used.

The same sort of cumulus clouds – with fairly vigorous updrafts and cloud-top temperatures of −5°C (23°F) or below – was also regarded as a candidate for the water droplet injection method. This was soon shown to be a very inefficient method because it required much more effort to spray water droplets upwards, against gravity, than to seed the clouds from above, and it was soon abandoned.

Australia is a country that could have benefited enormously from a successful programme of rain stimulation experiments; as well as the central desert there are vast tracts of semi-arid land which receive scanty and irregular rainfall, and which could be turned into more productive agricultural regions if it rained more frequently and in larger quantities. The Commonwealth Scientific and Industrial Research Organisation (CSIRO) set up a large and well-funded research programme in 1947 in order to determine whether or not cloud-seeding was the way forward, and if so, to establish which was the most efficient method. Successes were claimed, but once again it was virtually impossible to determine how much rain might have fallen anyway. What did become quite clear was that, however successful the seeding of individual clouds might have been, the huge investment of effort and expense produced marginal results. Furthermore, the theory of artificial rain-making held out very little hope of significantly increasing the amount and frequency of rain falling in a particular district, year in, year out. Consequently the research programme was scaled down after five years or so.

Conspiracy theories

The UK is a country where the artificial creation of additional rain would be regarded as an offence against common sense in most summers. Droughts do occur occasionally, of course, but most inhabitants of these green islands at the downwind end of the Atlantic Ocean think that, as a general rule, they get too much rain. Nevertheless, the UK Meteorological Office indulged in some desultory experimentation on medium-sized cumulus clouds during the late 1940s and early 1950s, with the same equivocal results as the Americans and Australians had achieved.

'The Day They Made it Rain' was the provocative and inaccurate title of a BBC radio programme which was broadcast in late August 2001. Its aim was to show that cloud-seeding experiments could have been responsible for the disastrous Lynmouth flood of 1952. This flood destroyed much of that north Devon village on the night of August 15–16, leaving 34 people drowned and 420 homeless.

The radio programme presented evidence of rain-making experiments in southern England in the early 1950s thanks to the detailed recollections of RAF personnel, then long retired. These airmen remembered meteorologists supervising the seeding of selected clouds, and the great satisfaction of the scientists when rain was subsequently reported. It was also claimed that these exercises were conducted in secret.

All this falls into the 'Never let the facts get in the way of a good story' category. Far from being secret, rain-making experiments were talked about all over the place in the early 1950s. The Royal Meteorological Society's popular magazine,

Weather, devoted a whole issue to the subject in July 1952 – just a month before the Lynmouth disaster. An article by E.G. Bowen described the systematic research programme of artificial stimulation of rainfall in Australia which had begun in 1947, while one of Britain's foremost atmospheric physicists of the day, Frank Ludlam of Imperial College, described in detail the physical processes underpinning cloud-seeding research in the UK.

As we have seen, any meteorologist with a rudimentary knowledge of cloud seeding could explain why it is preposterous to blame the Lynmouth flood on such experiments. The key to understanding why this is the case is that completely different rain-making processes are involved. Scientists involved in rainfall stimulation were only interested in seeding individual cumulus clouds – those cauliflower-shaped clouds, usually less than a mile across, which sometimes produce showers which may last 10 or 20 minutes. Injecting substantial amounts of dry ice or silver iodide into such a cloud stimulates the production of ice crystals in the cloud which in turn accelerates the rain-making process. But the cloud has to have sufficiently vigorous updrafts to spread the chemical throughout the cloud. This in turn means that the cloud may have eventually produced rain in any case, and the seeding merely caused it to happen earlier. As we have already seen, there has never been unequivocal evidence of how successful these rain-making programmes have been.

The storm which caused the 1952 disaster was not confined to the Lynmouth district. Heavy rain fell over the whole of south-west England and south Wales, and it was caused by a depression which had stagnated just to the south-west of the UK for two days. Similar depressions have triggered serious flooding in south-west England at regular intervals, and previous devastating floods hit Lynmouth in the eighteenth and nineteenth centuries. The August 1952 depression was over 500 kilometres (300 miles) across, and the prolonged heavy rain associated with it was caused by the large-scale lifting of very moist air. A fleet of RAF Hercules aircraft stuffed with dry ice would not have made a ha'porth of difference.

This very naughty radio programme gained a good deal of press coverage over subsequent days. It shows how simple it would be to stir up public feeling if weather-control experiments ever became demonstrably successful. It is only one short step from fomenting dissatisfaction among the populace to filling the courts with law-suits.

Chapter 18: Wartime weather wheezes

Smoke gets in your eyes

Winston Churchill became British Prime Minister on May 11 1940, and within a few weeks of taking office he was dabbling in rudimentary weather control. On May 29 he ordered the setting up of a working group to determine whether it was feasible to use smoke screens to hide military targets which were particularly vulnerable at night to bombing by enemy aircraft. It took just three weeks for the working party to come to a conclusion, for the establishment of a meteorological unit at the Ministry for Home Security, and for operational trials to be conducted. By July smoke screens were being used at ten vulnerable sites across England and Wales, and a team of meteorologists was employed to control how and when the smoke would be discharged.

Several of those young weather forecasters went on to become well-known figures in meteorology, including the unit's head, P.A. Sheppard, who subsequently became a director at the Meteorological Office and later Professor of Meteorology at Imperial College in London. Another, Dick Ogden, became head of the London Weather Centre in the 1970s, and this account draws substantially on his recollections.

During the early months of smoke screen operation it was agreed that deployment would only be required during periods when moonlight was bright enough to provide enemy aircraft with good visual identification of targets. Thus, for a week either side of full moon the forecasters had to predict all occasions when there was not a 100 per cent cloud-cover or thick fog, and also nights when the wind-speed was 20 mph or more, because it was impossible to produce an effective smoke screen if the wind was too strong. Once it was determined that a night was likely to be an operational one, they then had to forecast local wind conditions in considerable detail so that the smoke generators could be located to give the most effective cover to the building or site being protected. Ogden describes the deployment of smoke generators thus:

> Each consisted of a cylindrical pot containing 20 gallons (90 litres) or so of diesel oil, and surmounted by a tall chimney so that the whole assembly was about four feet (1.2 metres) high. Pots were positioned about ten yards (almost ten metres) apart, ideally in a semi-circle about 2000 to 3000 yards (roughly 2 to 3 km) upwind of the target. An alternative array for use in flat calm or very light and variable winds was that of a cross centred near the target and extending outwards to the circuit used when the wind was determinate. In practice, however, the ideal layout could rarely be achieved because generators had to be located on roads, or at least on tracks which could be used by lorries. The generators burnt about two gallons (nine litres) of diesel oil per hour and produced thick, black, oily smoke.

Later in the summer of 1940 a large-scale research programme into nocturnal wind patterns over urban and suburban land surfaces took place. Unlike pre-war research into winds which concentrated on the potential of high winds to cause damage to buildings, bridges, pylons, and forests, the greatest attention was now paid to occasions of very slight air movement. Much was learnt about the influence of river valleys and areas of rising ground, and also the differing effects of the 'surface roughness': airflow is more irregular and unpredictable over rougher surfaces such as urban areas compared with the relatively smooth surfaces found in open country. It also became evident that the generators had to be doused very quickly when natural radiation fog was about to form: thick fog polluted with oil droplets and heavy carbon particles resulted in visibility of five metres (about five yards) or less. Residual smoke and pollution often lingered into the following day. At Luton, in the south-east Midlands of England, the location of the Vauxhall motor-vehicle factory where thousands of military vehicles were built during the war, the number of mornings with fog reported during the winters of 1940–41 and 1941–42 was double that observed at an adjacent rural site, whereas the pre-war record showed only a small difference in fog frequency between the two locations.

The number of smoke screens increased progressively during 1941 and 1942, and at the climax of the programme more than 30 towns the length and breadth of England were involved. The winters of 1940–41 and 1941–42 presented another problem. Both were extremely cold and large parts of the UK were snow-covered for several weeks. The reflection of moonlight from the snow-cover provided ideal conditions for the accurate visual identification of targets by German raiders at night, so it was especially important during these periods to establish effective smoke screens. The location of smoke generators in residential areas was not exactly popular with the inhabitants, but the exigencies of war meant that people had to like it or lump it. The programme was finally wound up in September 1944.

FIDO: Bomber command's favourite pet

Dick Ogden was also intimately involved in the FIDO programme which began in 1942; FIDO officially stood for Fog

Investigation Dispersal Operation, but Ogden maintains that the origin of the acronym was actually 'Fog, Intensive Dispersal Of' which sounds plausible to anyone familiar with contemporary British military terminology. Fog and low cloud had been recognised as major hazards since the beginning of aviation, and attempts at artifical dissipation of fog had started shortly after the end of the Great War. Experimentation went down two separate routes: seeding and heating. American researchers investigated a variety of chemical dusts and solutions for spraying into the fog, notably calcium chloride which was regarded as being likely to provide suitable condensation nuclei. Not only were the trials unsuccessful, but calcium chloride is very corrosive, and in substantial quantities it is also toxic to plants and animals. Attempts were even made to spray electrically-charged sand into the fog, but this also failed to provide any worthwhile results.

In the UK the idea of burning fuel on airfields to supply heat was first considered in 1921. The physics was simple and depended on the fact that a given volume of air is able to hold more moisture when it is warm than when it is cold. Fog forms when slow-moving air cools at night to a point where it can no longer support its moisture, and the moisture condenses into water droplets. Supplying heat at ground level reverses the process, lifting the temperature sufficiently so that the air is no longer saturated. Water droplets in the fog are once again absorbed by the atmosphere in the form of water vapour.

Early analysis suggested that the quantity of fuel required was both uneconomic and inefficient and the idea was abandoned, and these findings were repeated when the subject was revisited in 1936–39, and again in 1941. The following year everything changed. Bomber Command under Air Marshal Arthur 'Bomber' Harris was now equipped with radar and making hundreds of successful night-time flights across Germany, but the number of aircraft lost to crashes in fog on the return was growing at an alarming rate. Lord Cherwell,

Churchill's scientific adviser, strongly encouraged his boss to get the Home Security Minister to set up some full-scale trials to determine whether fog dispersal by burning fuel was technically feasible, whatever the cost.

Ogden tells us that the first trial trial took place at Moody's Down, just north of Winchester, on November 4, 1942. An account by J.K. Gilchrist describes it thus:

> Two burner lines 200 yards long and 100 yards apart were lit. Between the rows, an 80-foot escape ladder borrowed from the Southampton Fire Service was erected, and up it went a local fireman to disappear – like the boy in the Indian rope trick – into the oblivion of the fog. Presently, however, as the burners got under way the fog began to lift and the fireman came into view; when the burners were turned down he disappeared again. FIDO was almost a fact.

The go-ahead to develop the system was authorized very rapidly by Churchill, and systems were established at 15 airfields from Carnaby, Yorkshire, in the north to Manston, Kent, in the south-east, and St Eval, Cornwall, in the south-west. The airfield at RAF Graveley, near Huntingdon, was the first where FIDO was operational and the first successful fog clearance was achieved on July 17, 1943. At five o'clock in the morning fog had closed in, the horizontal visibility falling to between 100 and 200 yards (90–180 metres); but within ten minutes of lighting the burners the fog had lifted and visibility was estimated at six miles (10 kilometres) – an astonishing success. Nevertheless, visibility was still bad as the aircraft descended towards the airfield and pilots had to use normal blind landing techniques on approach; they also had to contend with some unpleasant turbulence caused by the heated air just before touching down.

The consumption of fuel was enormous, and the cost of keeping airfields clear on foggy nights was staggering, but this was wartime and normal economic considerations were suspended for the duration. Most of the FIDO-equipped airfields burned about 320,000 litres (70,000 gallons) of petrol per hour, but four of the airfields had larger installations for a variety of reasons and these could burn as much as over a million litres (250,000 gallons) every hour. Thus a foggy night over England with all the airfields in use could have consumed something like 90 million litres (20 million gallons) of fuel. On one occasion the additional heat pumped into the atmosphere over Lincolnshire during FIDO operations triggered a thunderstorm.

Between 1943 and 1945 there were in all 161 successful fog clearances which permitted a total of 722 aircraft to land safely; there were also 182 FIDO-assisted take-offs, and a further 1700 FIDO-assisted landings in visibility that was poor but not officially within fog limits. The American air force bought into the scheme as they had not succeeded in developing their own system, and after D-Day a portable version of FIDO was developed for use by Allied pilots on continental airfields. Once the war had ended, however, economic and environmental considerations meant that the projected use of FIDO for civil aviation was not generally authorized, although one or two airfields maintained their installations until the late 1950s.

The smoke screens and the fog dispersal programme were both examples of deliberate small-scale modification of the weather. But it is interesting to note that the mechanism for each of these can also be observed in inadvertent modifications on a larger scale. Many European and American cities suffered from a semi-permanent winter smoke pall, resulting from industrial emissions until clean air legislation in the second half of the twentieth century reduced the quantity of carbon particles and sulphur dioxide being pumped into the urban atmosphere. In central London, for instance, December sunshine amounted to barely 25 per cent of the sunshine recorded in nearby rural areas during the late nineteenth and early twentieth centuries. Today, the urban atmosphere is much clearer, and the nocturnal heat emissions in large conurbations raises the temperature sufficiently during quiet winter weather to suppress the formation of fog. Thick fog is now reported in central London on fewer than half the days when it forms outside the built-up area, and December sunshine totals are 10 to 20 per cent higher in the centre compared with the outskirts.

Chapter 19: Man-made climate change

Introduction

Man-made climate change is not just so-called global warming. On a small scale human beings began to modify the climate of the district where they lived as soon as they started to cut down forests and drain marshland. The deforestation of Europe during the early centuries of the last millennium had a significant effect on the continent's climate, making it warmer and less humid in summer, colder in winter, and marginally drier in all seasons except possibly the summer. The same process has been taking place in North America during the last two centuries, and is under way at the moment in other parts of the world, notably in South America. The wholesale removal of forests in the shoulder of Brazil, inland from Recife and Fortaleza, during the twentieth century resulted in a quantifiable reduction in rainfall – a process known to geographers as desertification. But none of these modifications remotely approached the scale, in both space and time, or the predicted intensity, of the global changes which are now beginning to happen as a consequence of emissions of greenhouse gases into the atmosphere.

All the above examples could be described as accidental or inadvertent. There was no intention of changing the climate when the forests were being cut down, or when the industrial revolution began to change the chemical content of the atmosphere. But there are also examples of deliberate or intentional climate change, although – mercifully – most of these have remained on the drawing board. They include various *grands projets* of the leaders of the now defunct command economies of the last century, a number of hare-brained schemes of imaginative military men, as well as some crackpot notions suggested by eccentrics and conspiracy theorists. There are two big problems with deliberate climate change: our inability to undertake trial runs, and the legal consequences of any knock-on effects.

Until the 1970s most research into climate change concerned past climates. Then everything changed. Allegedly new theories (actually they were over 200 years old) were propounded about global warming and the so-called greenhouse effect resulting from the growing concentration of carbon dioxide and other gases in the atmosphere during the 1980s; these theories coincided with an increasing awareness in the liberal democracies of environmental issues, a resurgence in 'green' politics in many countries, and an expanding appetite for sensational stories in the media, so they were given huge publicity. More important, they also coincided with insistent signals from the climate record in all seven continents of the beginnings of a warming trend, so the subject caught the attention of a large number of scientists working not only in climatology and meteorology but in a variety of other allied fields.

Until the late 1980s speculation about changes in the climate were just that – speculation. Early computer models aimed at predicting future climate were grossly simplified and very crude. In 1973–74 one climate prediction project created a considerable stir when it suggested the strong possibility of a rapid decline into a new glacial episode during the twenty-first century, and H.C. Willett wrote a paper with the provocative title 'Do Recent Climatic Fluctuations Portend an Imminent Ice Age?' in the prominent scientific journal *Geofisica Internacional*. But several scientific papers were published around the same time which suggested that a major global warming episode was on the cards. For instance, the climatologist W.S.

Broecker wrote an article in August 1975 for the journal *Science* entitled 'Climate Change: Are we on the Brink of a Pronounced Global Warming?', while a year later the same journal published a paper by P.E. Damon and S. M. Kunen called 'Global Cooling? No. Southern Hemisphere Warming Trends may Indicate Onset of the CO_2 "Greenhouse" Effect'. A little later, in February 1978, the National Defense University of the USA – established and funded by the Pentagon – conducted a detailed survey of climate experts around the world to try to gauge the conventional wisdom of the time. There was little consensus, with an average expectation of a rise of 0.1°C (0.2°F) in global temperature between 1970 and 2000, but at one extreme they said that there was a 10 per cent chance of a 0.6°C (1.1°F) warming, and at the other a 10 per cent chance of a 0.3°C (0.3°F) cooling. All of the experts accepted that increasing levels of carbon dioxide in the Earth's atmosphere would tend to produce a global warming trend, but most of them believed that other factors such as volcanic activity and a natural cooling following the climatic 'mini optimum' of the early twentieth century would offset most or all of the warming.

Four important things happened between 1980 and 1990. First, the global cooling trend which had been evident since about 1940 ceased, and then went into reverse. Second, the revolution in computer technology allowed climatologists to assess the existing evidence of climate variability in a much less subjective way than hitherto. Third, those same powerful computers enabled the experts to build more and more sophisticated computer models which produced rather more useful predictions of how global and regional climates might change over the coming decades. And finally, ongoing research into the stratospheric ozone layer revealed that human activity was having a severely detrimental effect on ozone concentrations above the Antarctic continent during the spring and early summer seasons.

By the early 1990s climate experts found themselves in a completely new situation. Worldwide anxiety prompted by the latest computer predictions, exacerbated by sensationalist and irresponsible reporting in the print and broadcast media, the growing politicization of environmental issues and the increasing vociferousness of environmental pressure groups, the ambivalent and often conservative attitude of politicians of all shades of opinion who were unused to making decisions whose repercussions might not be felt for decades, and the vested interests who foresaw major financial losses if their particular industry were forced to change its production methods ... all these things thrust these scientists, who until then had been happy to potter around in their rather sleepy backwater, very firmly into the limelight.

Tensions exist between scientists and politicians, and will continue to exist for the foreseeable future. This is because governments need facts to enable them to make decisions, while scientists can only present facts about things that have happened in the past or are happening in the present. They make predictions about future developments to the best of their ability, aided by increasingly complex and sophisticated computer models, but these predictions will always be probabilities, not certainties. Rarely is a probability sufficient for the minister or secretary of state who would have to spend billions of pounds or dollars to act on it.

Global warming and the enhanced greenhouse effect

Human beings are able to live comfortably in the lowest few metres (few feet) of the atmosphere because they can rely on the climate there being more or less static. Temperature levels, in particular, change little when averaged over a period of years. This long-term thermal equilibrium is the result of a balance between heat energy coming from the sun, and outgoing energy which is radiated from the surface of the Earth into outer space. Much of the incoming shortwave radiation gets through to ground level because the atmosphere is transparent to it, but the outgoing radiation is at a much longer wavelength and the atmosphere is rather less transparent to this. In fact, some of the long wave radiation is absorbed by various gases in the air, in particular carbon dioxide, methane, ozone, and (especially) water vapour, and then re-radiated so that a proportion of it returns to the ground. In effect, then a certain amount of heat energy is trapped in the lower atmosphere, and this is often (erroneously) compared with the way heat builds up in a greenhouse on a sunny day – hence the popular term, the 'greenhouse effect'. In the same way, those gases in the air which absorb the energy are sometimes known as 'greenhouse gases'.

This natural 'greenhouse effect' means that we can live comfortably on the surface of our planet, because it maintains a global temperature at a very much higher level than would otherwise be the case. It is generally accepted that the natural greenhouse effect warms the globe by some 30 to 35°C (55 to 65°F). Without it, the mean global temperature would be 20 to 25°C below freezing and the Earth would be a completely uninhabitable ice-bound world.

Mankind cannot significantly change the quantity of water vapour – the most important of the greenhouse gases – that exists in the atmosphere, but we can change the proportion of carbon dioxide. Actually, we have been increasing the concentration of carbon dioxide in the air we breathe ever since the first humans cut down and burnt trees to make clearings in the forest. For millennia these changes in carbon dioxide levels were trivial, but since the beginning of the industrial revolution in Europe some 200 years ago there has been a steady and continuous rise in the proportion of carbon dioxide in the global atmosphere, so that CO_2 concentrations are now more than 25 per cent higher than they were in 1800. Carbon dioxide emissions have grown thanks to the burning of fossil fuels such as coal, natural gas, peat and oil in industrial processes and in motor vehicles. The same physical processes which explain why the Earth is warmer with carbon dioxide than it would be without, also explain why any increase in the amount of carbon dioxide in the atmosphere will cause an increase in temperature. Most climatologists are satisfied that this warming process is now detectable in the world's weather records, and that it will continue for the foreseeable future.

In the last 60 years we have also manufactured a completely new group of gases called chlorofluorcarbons or CFCs, for use as propellants in aerosol sprays, in packaging materials such as expanded polystyrene, and in commercial and domestic refrigeration units. These CFCs are also greenhouse gases, although they are better known for their damaging effect on the ozone layer in the stratosphere.

Industrial countries have been – and still are – the biggest contributors to the growth of greenhouse gases in the atmosphere. The USA emits ten times as much carbon dioxide as the UK or France or Italy, and one hundred times as much as Bangladesh. Put another way, the average US citizen contributes two to three times as much carbon dioxide as the average Briton or Australian, and 50 times as much as the average Bangladeshi. Emissions of greenhouse gases are no longer rising steadily, they are accelerating. At the present rate of industrial development around the world, carbon dioxide levels are predicted to double their 1800 concentration by the year 2025.

Most atmospheric physicists agree that a major growth in greenhouse gases in the atmosphere will result in a major rise in the overall average temperature of the entire Earth/atmosphere/ocean system. However, it is almost impossible – even with the latest computer models – to predict how the climate of a particular continent might change, still less what the effect would be on any individual country. As the temperature climbs, a number of feedback effects may come into play. For example, the warmer the atmosphere becomes, the more water vapour it can contain. The likely increase in cloud cover will divert some of the sun's radiation back into space, producing a cooling effect, but it must also be remembered that water vapour is itself a greenhouse gas and any increase in atmospheric moisture will therefore have a warming effect. Any change in the energy balance of our planet will amost certainly change the global wind patterns which in turn will change ocean currents, and some land areas are particularly vulnerable to such events. A retreat of ice and snow in the Arctic and Antarctic are also likely to create serious knock-on effects to both atmosphere and oceans, while the melting of the permafrost in Siberia, Alaska and Canada will release methane – another greenhouse gas – which has been locked in the ground for millions of years.

Global warming in the future

One of the good things that came out of the near panic which reigned in 1987–88 after the first serious predictions of global warming was the establishment of an international grouping of experts, eventually called the International Panel on Climate Change, or IPCC. These scientists were asked to co-ordinate research into the world's changing climate, and to deliver reports of the latest findings at regular intervals together with a consensus view in respect of the validity of the research.

The IPCC's first report was published in 1990, and subsequent ones have appeared roughly every five years. There has been no change in the broad view that there is a warming process going on in the Earth's atmosphere at the moment, that a substantial part of this process is the consequence of human activity, and that it will continue for the foreseeable future. A small minority of the panel's experts have expressed dissenting views about whether the warming signal can truly be distinguished from the underlying 'noise' in the climate record, and a few of them have also suggested that it is as yet impossible to conclude that any identifiable warming trend is due to the emission of greenhouse gases. The IPCC reports also summarise the results of a variety of long-term climate models run by climate researchers in several countries, all of which indicate a continued warming of the Earth's atmosphere (and therefore also the Earth's surface) during the coming century, although the rapidity of the global temperature change varies widely from one model to another. As knowledge continues to grow and computer models become more sophisticated, the predictions will no doubt change several times, possibly in different directions, but one would expect the results of the various models gradually to grow closer to each other.

Rising temperature is not the only important effect of global warming. Another is the gradual melting of the polar ice-caps, and in particular the possible destabilization and eventual collapse of the West Antarctic ice shelf. Were this to happen, enormous quantities of very cold and fresh water would be injected into the surface layers of the world's oceans, raising the sea level and perhaps causing sudden and dramatic changes in circulation patterns of both atmosphere and ocean. At the moment the balance of opinion is that the Antarctic ice shelves are safe for at least a century, and therefore global sea level is not expected to rise by more than half a metre (one to two feet) during the next half century. However, opinions may change as knowledge grows.

It is sometimes difficult to explain to the layman that global warming does not mean that all parts of the world will become progressively warmer. Far from it. A few regions may actually become colder, at least temporarily, while changes in rainfall patterns are likely to be very complex. In some regions, sections of the media have suggested that global warming is no bad thing because it would improve the local climate. In the UK, for instance, many newspaper features appeared in the early 1990s which suggested that Britain's climate would become rather more Mediterranean in character, with hot summers, mild winters, citrus orchards and vineyards, reduced heating bills, and teeming holiday beaches. They chose not to examine the down side of such a change – rising sea level resulting in beaches being submerged or washed away, flooded lowlands, pests and diseases hitherto unknown, summers plagued by photochemical smog, and large air-conditioning bills. Journalists also like to link any unusual weather event to global warming ... as if unusual weather events had never happened before.

There is much controversy about what we should do about the expected climate change during the twenty-first century. Environmental pressure groups and industrial vested interests vie for the attention of government ministers. It is difficult for our politicians to make balanced decisions when the scientists are unable to predict accurately the detail of those climatic changes.

The human race probably has the power to alter the global climate, but it has negligible control over the direction and magnitude of those changes. It is certainly possible that the computer models we have at the moment seriously over-estimate the warming trend due to increasing carbon dioxide levels, and if this were the case any changes forced on industrial conglomerates, which would undoubtedly be costly, disruptive, and politically sensitive, would also turn out to have been unnecessary. It is equally likely that the computer models underestimate the warming trend, and the consequences of that hardly bear thinking about. It is generally accepted that the atmosphere takes a long time – many decades – to respond to changes in emissions of greenhouse gases. This means that changes in the climate resulting from mankind's activities during the last half century are still in the pipeline, and cannot be stopped. The longer we put off the difficult decisions, the longer it will take to put things right.

Even now we can do some things to reduce greenhouse gas emissions that are worth doing anyway – for example, changing to renewable energy sources like wind power and solar power if and when they become economically competitive, discouraging the clearance of tropical rain forests, insulating buildings, and improving the efficiency of transportation systems. Moreover, increasing research spending to help develop those renewable sources of energy would cost comparatively little in the long run. Anyone who believes that no action need be taken should be reminded that the world population is doubling once every 40 to 45 years, so the emission of greenhouse gases will accelerate exponentially during future decades if we do not make any changes to the way we consume carbon-based fuel. And if it were to be proved in, say, 50 years time that it was vital to reduce these emissions, the problem would by then be several magnitudes greater.

Nevertheless, even if attempts at international agreement on restricting emissions of greenhouse gases, such as the Kyoto Protocol published in 1996, were to succeed, it would really only scratch at the surface of the problem. We really must put more effort into learning how to live in a warmer world where unexpected climatic fluctuations will periodically occur, and we must also aim at finding a variety of technological fixes to absorb or break down greenhouses gases at an economical cost, and ultimately to eliminate our reliance on carbon-based fuels. It is no good any more for oil companies to buy out inventions which might reduce our demand for oil, as they have been doing for half a century or more. And it is no good for environmentalists to bleat about the profligate use of oil and gas by rich western countries without being able to offer alternatives acceptable to these highly industrialized economies, and also to the rapidly growing economies of east and south Asia, South America and eastern Europe.

The climate of our home planet is constantly changing; it has always changed, and it will continue to do so. Climate change is, in part, a natural phenomenon. Year on year fluctuations are also part of the normal scheme of things. Therefore the prospect of an additional level of climate change in response to human intervention must be viewed in the context of those natural fluctuations. Nor must we be too quick to jump to conclusions. The global warming debate of the late 1980s took place against a background of a series of very hot summers in the USA which was probably pure coincidence, even though it helped focus attention on the message that the climatologists were trying to convey. Likewise, when a cold summer finally came along, it did not mean that global warming had somehow suddenly gone away.

In all parts of the world, gradual trends in the climate will be masked by much larger oscillations from year to year. Individual hot summers or warm winters are not, *per se*, evidence of a warming trend; likewise, cold snowy winters and cool rainy summers are not evidence of its absence. In the end, only time and the efforts of the climate experts, examining the statistics, calculating their averages, and running their computer models, will determine conclusively the degree to which anthropogenic global warming is happening.

Confusing ozone

Ozone is a confusing gas. The ozone layer in the high atmosphere is a 'good thing' because it protects us from harmful ultra-violet radiation coming from the sun. Ozone in the lowest layer of the atmosphere – where we live – is a 'bad thing' because it causes breathing difficulties and other health problems. The stratospheric ozone layer is a natural phenomenon, but low-level ozone is a pollutant resulting from the action of strong sunlight on the exhaust gases of motor vehicles.

To confuse things further, ozone is actually a form of oxygen – the most vital life-sustaining constituent of the atmosphere. But it is a comparatively rare sort of oxygen, comprising three atoms of the element instead of the usual two. Thus it is normally referred to in shorthand as O_3, whereas the regular oxygen is written O_2. Roughly 20 per cent of the atmosphere is oxygen, but ozone forms less than one millionth part of it, and nearly all of that is in the high atmosphere, some 50 to 65 kilometres (30 to 40 miles) up.

The most valuable task performed by the stratospheric ozone layer is to filter out much of the ultra-violet (UV) radiation which comes from the sun. It prevents very nearly all the very damaging UV-C wavelengths from getting through to the ground, and it removes some 80 per cent of the less damaging UV-B wavelengths. Ultra-violet radiation accelerates the wrinkling and ageing of our skin, it causes sunburn, it contributes to snow blindness, it can also cause damage to our eyes, and it is responsible for some melanomas (skin cancers). Indeed, in sunny climates colonized by white-skinned peoples such as California, South Africa and Australia, between 25 and 50 per cent of all cancer deaths amongst Caucasians are from melanomas induced by UV-C radiation.

The ozone layer is produced by the effect of sunlight on ordinary oxygen, and in the course of time it again breaks down into the more usual diatomic oxygen. Ozone can also be broken down by other chemicals that are found naturally in the atmosphere, while artificial chemicals can do the same. Indeed, chlorofluorocarbons – CFCs – scavenge ozone voraciously under certain circumstances, particularly under the influence of sunlight when the temperature is below −60°C (−76°F).

The stratospheric ozone will only remain to protect us from damaging ultraviolet rays as long as there is a balance between the generation and destruction of the gas. Ozone depletion has been measured above the Antarctic since the 1970s, and it is at its greatest during spring. This depletion accelerated during the 1980s and early 1990s. At first it seemed that conditions in the northern hemisphere were not quite right for the development of an 'ozone hole' above the Arctic, but observations showed substantial loss of the gas above northern Europe during the early springs of 1991 and 1996.

Governments reacted surprisingly quickly to this particular threat, setting up a panel of experts as early as 1977, and an international plan known as the Montreal Protocol was agreed in 1985. The most important result of this has been the rapid phasing out of CFCs. However, the atmosphere responds very slowly to these changes, and although there are signs that ozone depletion over the Antarctic levelled out by the year 2000, concentrations of the gas are not expected to return to 1980 levels until after 2050.

Chapter 20: Deliberate modification of the climate

Introduction

Intentional modification of the climate on a local scale dates back to prehistoric times when farmers in the Middle East and around the shores of the Mediterranean Sea planted lines of trees to create shelter belts to reduce wind speed, protect crops, and prevent dry soils from being blown away. One area where they were, and still are, particularly effective is the Rhône Valley in south-east France where the *mistral* wind is such a prominent feature of the local climate. Shelter belts are now found throughout the world wherever a prevailing wind presents problems for agriculture.

Reversing the Rivers

At the other end of the spectrum, an extraordinary scheme to alter the drainage patterns of a large swath of central Asia covering over 10 million square kilometres (4 million square miles) was planned by Soviet leaders in the 1950s and 1960s.

The growth of urban population in what was then Soviet Central Asia, primarily in Turkmenistan, Uzbekistan, Kazakhstan, and adjacent parts of Russia, had been putting increasing pressure on the supply of water in the region as early as the 1920s. To feed the expanding population, poor grazing land was converted to growing crops, chiefly wheat, maize, rye and cotton, but rainfall in the region is so scanty that irrigation was introduced across vast areas, using water from rivers flowing into the Caspian and Aral Seas. Irrigation is an inefficient use of a scarce resource, and a large proportion of the water used in this way evaporates, especially in regions like this where the atmosphere is excessively dry and the climate often sunny and windy. By 1950 water levels had fallen sharply, particularly in the Aral Sea which had shrunk by 25 per cent compared with its size at the beginning of the century. At the present rate of shrinkage, it will have vanished completely by 2030.

The plan, nicknamed 'The Reversing of the Rivers', was devised with the encouragement of the Soviet leader Nikita Kruschev during the late 1950s, when more and more land on the margins of cultivation were being opened up for agriculture. It was an extremely ambitious scheme and depended on the fact that some 70 per cent of the Soviet Union is drained by rivers flowing northwards to the Arctic Ocean. It was proposed that several of those rivers, including the Ob and Yenesei, should be dammed, allowing the headwaters to be directed along thousands of kilometres (miles) of artificially constructed channels flowing from east to west, finally disgorging into the Aral and Caspian Seas as well as a series of huge new reservoirs in the Kazakhstan lowlands. At the same time the vast area of marshland covering over 600,000 square kilometres (about 250,000 square miles) of north-western Siberia could be drained, while additional ideas included the diversion of Arctic rivers and the draining of marshland to the west of the Urals into the catchments of the rivers Volga and

Don to provide additional water resources to industry in southern Russia. It was envisaged that this monumental engineering project would take at least 30 years to complete.

Clearly, the draining of extensive marshland and the watering of the semi-arid lands of central Asia would change the climate on a local scale. But it was recognized right from the start that the removal of the bulk of the fresh water flowing into the Arctic Ocean could have much wider climatic repercussions. It seems that the Soviet government was little bothered about how climatic change might affect other countries, but its own climatologists warned that a serious global change was very likely which would of course have affected the Soviet Union as much as any other nation.

The only other rivers delivering significant amounts of fresh water to the Arctic Ocean are the Lena in north-eastern Siberia and the Mackenzie in Canada; the Soviet plan could have reduced the volume of fresh water reaching the ocean by 30 to 40 per cent. Soviet models showed that this fresh water spreads out across the surface of the Arctic, and freezes easily during the winter. Remember that water with little or no salt content is buoyant and freezes at 0°C (32°F); salt water freezes at about −2°C (28°F) but at this temperature it is not buoyant, so overturning occurs through a considerable depth of water, preventing freezing until a substantial layer of ocean has been cooled to that temperature. Thus, they reasoned, removing the fresh water supply could easily result in much of the Arctic Ocean being ice-free throughout the year, with nothing more than coastal ice shelves developing during the winter season.

The resulting change in climate was predicted to be catastrophic. It was estimated that the mean annual temperature in the Arctic Basin, including Soviet island groups such as Novaya Zemlya and Severnaya Zemlya, would rise by more than 20°C (36°F). The chief middle-latitude depression track over the Atlantic Ocean would extend through the Norwegian Sea into the Russian Arctic, while more Pacific Ocean depressions would cross the far east of Siberia and also western Alaska to reach the opposite end of the Arctic Ocean. This theory indicated that a consequence would be a drop in rainfall over much of European Russia, perhaps in central Asia too – the very region the scheme was designed to help. More worrying, there would probably also be a significant drying of western Europe, and western parts of Canada and the USA, and it was reasoned that these countries were hardly likely to accept a major change in their climates without complaint.

The original plans were therefore abandoned, but a revised and less grandiose proposal resurfaced in the 1980s, suggesting a much more limited diversion of Siberian water into the Aral/Caspian region. Climate computer models in Canada and the UK were run in the mid-1990s on the basis of reduced fresh water input into the Arctic Ocean, and the results of these indicated that the Soviet climatologists of the 1950s and 1960s had been unduly pessimistic. However, these models are still not good enough for us to take their results as gospel, and in any case the political upheavals in the former Soviet Union make it virtually impossible now to contemplate – either financially or politically – what would now be an international engineering scheme of unprecedented size.

Our ability to change the climate of our planet is very limited, but it is all the more dangerous for that. The quantity of energy we can generate is trivial compared with the storms and jetstreams, the heat and the electricity, which are merely parts of the great weather machine. Where mankind does have an influence is at the margins: in those areas where we can tip the balance and generate a chain-reaction. It is like a child in one of those television cartoons encountering a towering display of tins in a supermarket. We all know the child is going to remove one of the tins from the bottom of the display, nothing will happen for a few seconds, the tower then teeters and finally collapses, burying the poor infant. By messing with the composition of the atmosphere, or diverting rivers, or melting polar ice, we are pulling out one of those tins. And remember, it takes an awfully long time to rebuild the tower.

SECTION FOUR

The future

Chapter 21: Forecasting the weather in the twenty-first century

Trying to predict advances in meteorology during this century is probably more foolhardy than forecasting the weather next year. The accuracy of both exercises will probably be similar. However, some of the developments which will reach fruition during the next decade or two are already under way, although even the next 20 years will probably spring a few surprises as well. Beyond that ... well, it is simply speculation.

If anyone needs convincing of this, let them look back 50 or 100 years, and examine what the main concerns of meteorologists were then. We might even find in the literature some idea of our predecessors' speculation as to the future of meteorology.

In fact we are very lucky because the Royal Meteorological Society marked its golden jubilee in 1900 and its centenary in 1950 and, as usually happens on such occasions, the respective presidents of the society made wide-ranging lectures to the assembled fellows.

Predictions from the past

The centrepiece of the jubilee celebration was the address given by Dr Theodore Williams, the society's president, followed by contributions from several other eminent meteorologists. One of these was Napier Shaw who was to become one of the most successful directors of the UK Meteorological

Office, and he turned his attention to likely developments during the twentieth century. He foresaw ever more people involved in data gathering, and the importance of a central office to co-ordinate the collection of information. He expected technological advances which would bring much more data from the upper atmosphere, a region he regarded as vital for any big improvement in the quality of weather forecasts. Shaw also called for a greater contribution by physicists, mathematicians and chemists to develop a much more scientific understanding of how the atmosphere worked, and he predicted that many long-held shibboleths would vanish as more rigorous scientific procedures displaced the empirical rules and descriptive analyses which at the end of the nineteenth century dominated meteorology. Other contributors foresaw a rapid improvement in short-term weather forecasts, and one suggested that 'the more observations we make, the more factual information we examine, the closer comes the day when accurate long-term forecasts will be made'. One hundred years later we are still waiting.

The dominance of description over scientific research can be seen in the titles of papers in the *Quarterly Journal* of the Royal Meteorological Society that year. The biggest contribution of all, filling 58 pages and including 21 illustrations, was entitled 'The climatic conditions necessary for the propogation and spread of the plague'. Others titles included 'a remarkable dust haze experienced at Tenerife in the Canary islands', 'the circulation of water in the north Atlantic', 'remarks on the weather conditions of the steamship track between Fiji and Hawaii', and 'the Wiltshire whirlwind of October 1, 1899'.

The year 1900 also saw an International Meteorological Congress in Paris, and the main topics of discussion there were 'terrestrial magnetism and atmospheric electricity', 'aeronautics', 'clouds', and 'radiation and insolation'.

The hundredth birthday celebrations were built around a series of open meetings which addressed the issues of the day.

These included: 'radiation and its effect on the lower atmosphere', 'the physics of clouds and precipitation', 'the structure of weather systems', 'the general circulation of the atmosphere', 'climate change', and 'meteorology in the community'. Several of these topics were on the agenda 50 years before, and none would be out of place in a meteorological symposium 50 years later, notably climate change, although the subject matter would be appreciably different.

The *Quarterly Journal* of the Royal Meteorological Society, one of the leading research publications, contained cutting-edge reports on the internal workings of clouds, the introduction of upper-air data into weather forecasting process, the development of radar as a meteorological tool, and the measurement and evaluation of evaporation. Along with these pretty abstruse papers there were some relatively simple descriptive notes on, for instance, the microclimatology of a potato crop, the variation in time and space of thunderstorm severity, and climatic fluctuations of the preceding hundred years.

Sir Robert Watson-Watt made some predictions in his presidential address during that centenary year. He believed there would be a large increase in the number of people working in meteorology and climatology by the end of the twentieth century, not only in national meteorological services but also in research institutions and in the private sector. He expected a greater proportion of these people to be involved in research as against the preparation of forecasts and the collection of observations.

Elsewhere we learn that the meteorological services were lobbying hard for investment in numerical weather prediction, claiming that 'accurate weather forecasts for several days ahead' would quickly follow. Long-range forecasting was also considered by some to be a nut waiting to be cracked, and that co-ordinated research projects would probably lead to successful 30-day and 90-day forecasts by the 1970s. Fifty years later we are still waiting.

Computers were regarded as a welcome development which would 'eventually free meteorologists of most of the mundane clerical tasks – the filling of ledgers, the calculations of averages, and such like'. Money was always in short supply, but it was not regarded as ever likely to be a major problem, and adequate government funding for national meteorological services and for university research was expected to continue into the far future. Meteorological services to industry, commerce and the general public were mainly provided by a rapidly growing private sector in the USA, but in most other countries they all came under the roof of the national meteorological office. Because the work was centrally funded the cost of these services (except in the USA) was minimal, and no-one foresaw any significant change in this arrangement in the decades ahead.

The predictions made in 1900 and 1950 were not at all bad, so far as they went. Napier Shaw was his usual perspicacious self, and Robert Watson-Watt hit a couple of nails on the head too. The meteorological holy grail of long-range forecasting figured prominently on the wish lists in both 1900 and 1950, but even by 2000 progess in this field has been very modest.

It was what Shaw, Watson-Watt and the others failed to predict that was telling. None of them could have had the remotest inkling of the power and versatility that computers have now, or of the complexity of the numerical prediction models that we now use, or of the expense of running modern forecasting offices. They would have been disappointed that improvements in weather forecasting, impressive as they are, did not come more easily. They would be horrified at the man-made changes to the world's climate which seem now to be accelerating, and yet excited and motivated by the possibility that meteorologists and climatologists would be at the forefront of research into understanding the problem. And they would be perplexed beyond belief by the tight financial framework within which meteorology now has to operate, and that users of meteorological services now have to pay large amounts of money for information which was – in their day – virtually free.

Indeed, the public service ethos which pervaded all national meteorological institutions until the last quarter of the twentieth century has been severely eroded during the last 25 years or so; were Shaw and Watson-Watt able to travel through time to the present day they would struggle – and perhaps fail – to come to terms with the rampant commercialism which characterizes meteorology now.

So what does all this tell us about what our meteorological heirs and successors may be doing in 50 or 100 years time? We can perhaps make the following broad predictions. The leading academic scientists will still be working in many of the same fields as present-day researchers, for example in the effect of solar radiation on the atmosphere, in the physics and chemistry of clouds and thunderstorms, and in the field of climate change; but there will be many new areas of research.

Some important developments during the next 20 or 30 years can probably be extrapolated from the present day, but our particular scientific discipline can be expected to enter many new and possibly unforeseen avenues. The relative importance of meteorology and climatology will almost certainly change, while we can only speculate wildly about the rate of advance in remote sensing of the atmosphere, NWP models, and computer capabilities, and the sort of changes which may occur in the dissemination of information to customers and to the public at large, and in the financing of meteorological institutions.

There will probably be some surprises as well, perhaps nasty ones, consequent upon changes elsewhere in society – political, social, monetary – which we can only guess at.

Extrapolating into the twenty-first Century

The improvements in the accuracy of forecasts 24, 48, 72 and 96 hours ahead during the last 30 years have been erratic, but the general improving trend has shown no real signs of slowing. At the same rate we can expect a further 50 per cent improvement in forecasts for the next day by 2030. Predictions for two days ahead should be better than 24-hour forecasts are now, and those for four days ahead should be more accurate than 24-hour forecasts were in 1980.

These advances should come as the result of work which is already under way in specific areas of research. Arguably the most important of these will result in the inclusion of meteorological processes which take place on a scale which is at present too small for incorporation in NWP models, such as local variations of incoming and outgoing radiation from the Earth's surface, the formation and dissipation of clouds, and the small-scale characteristics of certain rain and snow systems. Almost as important are the expected advances in measuring the atmosphere in the coming decades: remote sensors on satellites, on aircraft, at sea, and on land, will provide data which is more reliable, more accurate, more frequent, and which covers much more of the atmosphere.

Of course we will need an exponential growth in computer power to handle all this additional data and to run the much more complex forecasting models, and the computing industry promises us this as well. In the year 2000 the super-computers used by national meteorological services could handle something like 100 billion caculations per second, but by 2030 it is estimated that the equivalent computers may be a billion times faster even than this. As well as coping with a rapid growth in the quantity of data and the complexity of models, such computing power would also permit a much greater use of *ensemble* forecasts, which would be of particular use in determining what factors cause weather forecasts to go wrong. This in turn should pinpoint areas for future research.

Nothwithstanding these advances, virtually all meteorologists accept that numerical weather prediction cannot continue getting more accurate indefinitely. In other words, there is a theoretical limit to the predictability of the atmosphere. Many scientists believe we will approach that limit during the twenty-first century, perhaps as early as 2050, and that the signs of its approach may soon become apparent with huge investments of time and money producing smaller and smaller improvements. In any case, before that theoretical limit is reached it will become economically unviable to pump more and more resources into numerical forecasting for a progressively smaller return.

It is also likely that the part played by human meteorologists will continue to diminish. A supervisory role will probably be the limit of his or her ambitions in 20 years time. The requirement of human forecasters to intervene in the event of a catastrophic systems failure seems likely to atrophy as back-up systems become more and more common. Some forecasters think that there will always be a job for them in interpreting the computer output and tailoring the forecast message to the particular requirements of the customer, but they are probably whistling in the wind. Even this role is already being eroded: airline pilots have already been self-briefing for ten years, and more and more users of specialized forecasts prefer to receive their service by computer, notwithstanding the fact that many over-40s are barely computer literate. In the future, when using computers is second nature to all, one would not expect many users of weather information to prefer a personal service to self-briefing.

The splintering of radio and television broadcast services predicted for the first half of the new century may yet kill off the highly paid and occasionally iconic national weather presenter. Although there may be an increase in the requirement for talking heads on the increasingly narrow-cast television/webcast services of the future, it is unlikely that the occupants of these posts will be either as famous or as well-paid as they are just now. In any case, work is already well in hand to create 'virtual' presenters, and to provide them with scripts compiled without human intervention. Weather information is fed into a text generator which formats the information into phrases and sentences characteristically found in weather forecasts. Such so-called 'concatenated' scripts are still clumsy, and sometimes laughable, but they are improving. One can easily imagine this becoming the norm on television, radio, mobile phones and the internet by 2020, and possibly as early as 2010.

A few more technological steps forward could take us into a home where a virtual presenter of your choice in the form of a hologram can be programmed to deliver a weather forecast tailored to your particular requirements, or to answer questions about the weather at given places and for given time periods. Such advances may seem exciting from our perspective, but it is quite possible that they will be seen as little more than gimmicks in a society where efficient self-briefing may well be the norm when it comes to acquiring daily information.

Always expect the unexpected

The progressive improvement in the quality of weather forecasts covering the next few days may have some surprising repercussions. In the late twentieth century weather forecasts were often disbelieved and the forecasters themselves derided, even though the forecast accuracy was improving. Now at the beginning of the twenty-first century a substantial majority of people believe that weather predictions are accurate, and they sometimes make important decisions based on an assumption of reliability, although they may still be loath to praise forecasters publicly. Their confidence in the science may also be periodically dented because they have used a forecast which was out of date, or over-generalized, or poorly presented. Given further improvements in forecast accuracy on the one hand, and the quality and currency of the product disseminated to the general public on the other, we may well reach a point in, say, 30 or 40 years time when weather predictions for 24 and 48 hours ahead are routinely regarded as 'fact', and incorrect forecasts are so infrequent as to be generally discounted.

One of the core public services provided by national meteorological offices as we enter the twenty-first century is the provision of warnings of severe or destructive weather events. This is described as a 'public' service because it is regarded as essential to the economic and social well-being of the community, minimizing the loss of lives and property; it is also a free service, or more correctly one paid for by the taxpayer out of public funds. As forecasts grow in accuracy we can expect the public attitude to these severe weather warnings to change.

In countries with authoritarian régimes the political infrastructure already allows for the forced evacuation of cities and regions when a disaster, a typhoon or hurricane, for example, is predicted. In practice, however, such evacuations are rarely enforced because the logistical problems of uprooting a large population for an indefinite period of time are too great. Even in the liberal democracies of the western world, people may be forcibly removed from small geographical areas in a serious emergency, where a leak of toxic gas has occurred or where a terrorist bomb has been planted. In London plans existed to evacuate millions of people from the floodplain of the Thames, in the event of an imminent tidal surge from the North Sea into the Thames Estuary, thence upstream into the capital. Over 250 square kilometres (about 100 square miles) of the urban area was regarded as 'at risk', including the national financial and political centres in the City of London and Westminster respectively. These plans were instituted after the great east coast floods of 1953, and remained in force until the completion of the Thames Barrier at Woolwich in 1983.

However, warnings of severe weather such as hurricanes or cyclones typically bring forth 'advice' or 'encouragement' rather than 'instructions'. Along those heavily populated parts of the American coastline which are subject to occasional visits by hurricanes – Texas, Louisiana, Florida, the Carolinas – people are advised to move to a designated hurricane shelter rather than flee the region; once again, the logistical difficulty of moving a large population out of a limited geographical area by road within a short timescale is prohibitive. Nevertheless, many do choose to leave, packing their cars with family and belongings, and resulting in gridlock on the road system which in turn may threaten the lives of those caught up in the congestion if the arrival of the storm is imminent.

A longer warning interval, an adequate network of designated shelters well inland, and, it has to be said, the political will, may yet see some governments ordering large-scale evacuations when a hurricane, typhoon or cyclone is predicted to make landfall. India and Bangladesh are the two countries which would benefit most from such a development, for they bore the greatest proportion of lives lost to such storms during the twentieth century. The November 1970 cyclone alone cost Bangladesh (East Pakistan as it then was) an estimated 400,000 casualties. But China, the Philippines, Taiwan, Vietnam, the Caribbean islands, Mexico, and the countries of Central America have also suffered considerable loss of life to these tropical storms at irregular intervals in the past. Americans and Australians of the present day may baulk at the idea of legally enforcible evacuation, but an extended track record of accurate warnings together with a slight change in attitude on behalf of the public at large could yet see it happen even in those two individualistic nations. Rather than physically removing people who choose to ignore such a law, one envisages instead a legal penalty being enforced.

Perhaps more likely would be a legal requirement for homes and commercial buildings in those at-risk parts of the US coastline to be built to a higher specification than elsewhere, with heavy duty shutters to protect windows and doors, and a 'hurricane shelter room' where stocks of water and food would be kept. A major hurricane disaster in, for example, Florida in the next few years could result in a change in the present *laissez-faire* attitude to building on the coastal fringes where the tidal surge associated with a hurricane can do immense damage. Such a circumstance could possibly mean the outlawing of new house construction in such places; rather than an outright legal sanction it is more probable that household insurance would be made an enforcible legal requirement, and with premiums in those threatened coastal zones becoming prohibitively large, few if any new homes would be built there. Similarly, we might envisage a legal requirement for all homes to have a tornado shelter in those parts of the USA where tornadoes are most frequent – in Texas, the Great Plains, the mid-West, and the south-eastern states, for example.

In Britain the floods of autumn and winter 2000–2001 provided a timely reminder of the dangers of building extensively on the flood plains of rivers, of the lack of investment over many decades in flood defence schemes, and of the subtle corruption of local politicians. Many examples were revealed of planning decisions being overturned by local authorities, allowing those authorities to gain financially by permitting developers to build new homes in areas at risk from flooding. Thousands of families endured repeated bouts of flooding during that season, the wettest for over a century. The misery was revisited the following year when many of them discovered that insurance companies were no longer prepared to insure their properties against future flooding. Britain's Environment Agency has also drawn up detailed maps of those parts of the country where river and coastal flooding are significant threats. Here we see the market already intervening to change attitudes. The identification of flood-risk zones and the withdrawal of insurance cover will effectively end the explosion of speculative building on flood-plains which began in the 1960s.

On the whole Britain's weather is relatively benign, and individual weather events which threaten the lives of the people are mercifully rare. However, destructive windstorms do occur in most winters, and occasionally there is considerable loss of life. One such gale in October 1987 killed 18 people, and another in January 1990 killed 48. Given the accurate and timely weather forecasts which we believe will be in place before the middle of the new century, it is quite feasible that a future British government will consider issuing a 'stay at home' order during such events, although one doubts that it would carry the full weight of an enforced curfew. The British people are never slow to seize an opportunity not to go to work, and if the government encourages them to stay at home they will not need telling twice. Such regulations could also be used on those rare occasions when other types of adverse weather may dislocate normal day-to-day activity: snowstorms, dense fog, and severe pollution episodes come to mind.

Even in that most litigious of all countries, the USA, lawsuits taken out against weather forecasters for 'getting it wrong' have, to date, almost all been thrown out. An inaccurate forecast is not regarded as sufficient grounds for a successful legal action; a forecast is, after all, just that, and not a statement of fact. It is generally required to prove negligence on the part of the forecasting organization or of an individual forecaster, and negligence is notoriously difficult to prove in cases involving scientific judgement. It is, of course, unwise to try to second guess how the various legal systems around the world will react to such cases in the future, and it is probable that a few court actions against meteorologists may be won by the plaintiffs. However, given the growth in automation, the installation of additional safeguards, and the increasing accuracy of the forecasts, it seems likely that it will actually become even more difficult to prove negligence in the future.

Insurance is a different matter: not insurance against the weather, but against a failed forecast. In a world where 24 hour forecasts are, say, accurate on at least 95 per cent of occasions – this could well be the state of play by 2050 – some insurance companies may offer insurance against personal or commercial losses suffered when decisions are made on the basis of a weather forecast which turns out to be wrong. For instance, a day's play at the All-England Tennis Championships at Wimbledon could be postponed because the forecast is for continuous rain on that day. The Wimbledon authorities take their decision in order to minimize the disruption to spectators, players, and, not least, themselves. In the event the weather stays dry, and the All-England Club are recompensed for the unnecessary reorganization by the insurance company. This is a natural extension of the 'pluvius' insurance policies which already exist, whereby outdoor events which are planned and organized long in advance insure themselves against financial loss if rain spoils the day. In a changeable climate like Britain's, the premiums are high and the payments are not great, because rain is rarely far away. But if the insurance is taken out against a prediction which, on average, goes wrong on only one occasion in 20 (and by no means all of those will involve rain falling on a predicted dry day, or vice versa), a relatively small premium could guarantee a substantial compensation.

We can speculate briefly – and undoubtedly inaccurately – on the sort of information delivery systems which might be in place by the middle of the century. Many people already carry personal information sources such as personal organizers and palm-top computers, and these can easily be plugged into mobile phones to provide a link to the internet. It does not require a leap of imagination to envisage fully integrated voice-operated ultra-microcomputers, permanently linked to whatever the internet has become by then, maybe strapped to the wrist, folded or rolled up in a pocket, or incorporated in a pair of spectacles or sunglasses. There would be no need to monitor the weather ourselves; rather, we would be automatically linked to the nearest site in a network of weather monitoring stations around the country which would transmit to us the current ambient measurements of temperature, humidity, wind speed and direction, solar radiation, or whatever other parameters we may have asked for. Our personal computer could be programmed to be proactive, for example informing us when the humidity is approaching a level that our personal virtual physician has deemed to be unhealthy or merely uncomfortable, or telling us when rain is five minutes away.

Some of us already spend part of the time in our own artificial microclimates, with homes and cars benefiting from central heating systems, air conditioners, and humidifiers. It is quite possible that at some point in the not too distant future we will wear clothes which are able to maintain optimum personal comfort by adjusting the temperature or humidity of our personal space in response to changes in the weather. It is very likely that we will spend less and less time in the real atmosphere, experiencing real weather. One could even imagine 'extreme weather holidays' becoming a popular leisure pursuit.

Chapter 22: The new climatology

Climatology has until recently been a backwater: a dusty old discipline peopled by crusty old boffins quietly going about their business and absorbing comparatively small amounts of money in the form of research grants. The threat of global climate change has altered all that. Climatologists were suddenly thrust, blinking furiously, into the media spotlight in the late 1980s and 1990s, were courted by presidents and prime ministers, and found that the financial floodgates had been opened.

The new century has delivered a new climatology. Thrusting young scientists coming up through university now see it as one of the sexy sciences, overflowing with research money and with countless opportunities to appear on television and to rub shoulders with leading politicians. Whether that is a good thing or not remains to be seen.

In any case, climatology will continue to grow as long as the warming trend in the world's climate continues. One surprising development may be the reversal in importance of climatology and meteorology compared with the pre-2000 status quo. We have already noted that it is quite possible that the theoretical limit to the predictability of the atmosphere may be reached before the middle of the century, with the accuracy of weather forecasts reaching a plateau. We have also seen that production and dissemination of forecasts will become largely automated during the next two or three decades, and that operational meteorologists will probably become supervisory technicians. If that barrier to predictability is reached the sources of investment in new meteorological research would quickly dry up; meteorology could then become the poor relation, the backwater.

Climate is actually a much bigger and more intractable problem than weather, and it has the potential to absorb much more money and effort in seeking answers to the problems it poses. Weather forecasting is simply a matter of applying known physical laws to a given state of the atmosphere in order to predict its behaviour over a relatively short period of time. This approach cannot be used for predicting long-term changes in the atmosphere. At the moment climate prediction models are essentially empirical; that is, they assess changes in what are called 'forcing mechanisms' – things like the sea-surface temperature, the concentrations of various gases in the atmosphere, the output of the sun, the various astronomical cycles – and predict the effect that these changes will have on the atmosphere's behaviour based on several broad assumptions. It is likely that improvements in climate forecasting will continue on a slow and painstaking path for several decades, those assumptions being fine-tuned, new relationships being found between the forcing mechanisms and the response of the atmosphere, and new forcing factors being discovered. Meteorologists learn about how good their weather prediction models are every day by comparing the forecast with the reality; clearly, such feedback is not available to climatologists whose predictions may cover the next hundred years or more.

It is almost axiomatic from our viewpoint, little more than ten years into an internationally co-ordinated attempt to understand global climate change, that the picture will become clearer with time. More data, more research, and better climate models, it is widely believed, will clear the view and before long we will understand whether the warming trend is serious, how long-lasting it will be, what proportion of it is down to industrial activity and what proportion can be explained by natural changes in the climate. We should not be so optimistic. As long as climate models are wholly or partly empirical, the knowledge we gain from them will remain equivocal. And there will always be plenty of people around the fringes of the science with political and commercial axes to grind, who will continue to generate a lot of noise and heat while shedding very little light. As long as there are vested interests involved in the debate on climate change, the waters will remain muddy.

The resources directed at climatological research during the coming century may, as a spin off, provide some impetus to the development of useful long-range weather forecasts – that is, forecasts for one, three, or even twelve months ahead. Some advances are therefore likely in this field, but real improvements may be as elusive for us and our successors as they were for Napier Shaw and Robert Watson-Watt. And one thing almost goes without saying: the prospect of forecasting for a particular place and a particular date more than, say, two weeks ahead will remain the domain of the eccentric, the deluded, and the rogue.

Chapter 23: Can we make a forecast for the twenty-first century?

Introduction

It is essential to keep in mind that the so-called predictions provided by climate models are limited by four things: first, the necessary simplification of the atmosphere involved in creating the model; second, the assumptions made about influences (including human activity) external to the climate system; third, the accuracy of the data being input; and fourth, our inability to see how well the model is working. After all, that is what a model is – a version of reality, usually smaller and simpler than the original.

Numerical weather prediction models simplify the atmosphere to produce a forecast of the weather for the next few days. In the same way, climate models which aim to predict broadscale changes in global and regional climatic conditions over many decades or even centuries also use a simplified representation of the atmosphere and its interaction with the oceans and continents. There are several limiting factors here which require a reduction of complexity. These include the time it takes to run the model, the finite number of observations of the state of the atmosphere in real time (because we are unable to measure the temperature, pressure and humidity at every point in the atmosphere and on the Earth's surface), and the imperfection of our knowledge in respect of the physical processes at work in the air around us. In particular, interactions between the atmosphere and the oceans is still poorly understood.

Assumptions about external influences on the atmosphere are crucial when a model is created to examine the probable impact on the climate of changes in those influences. For instance, we do not know with any accuracy the quantity of greenhouse gases that will be released by human activity in 10, 20, 50 or 100 years time. Moreover, we do not know what other greenhouse gases presently locked up in the oceans, the ice-caps and the tundra, may find their way into the atmosphere as the global climate changes. And we can only estimate what proportion of those gases might be re-absorbed into the oceans and the forests. Thus the same climate model might be run several times with different assumptions, producing different results each time, and each one of those results is only a prediction and not a guarantee of reality.

Any computer model which aims to predict the weather of the next few days or the climate of the next few decades requires starting conditions. Modern weather prediction models no longer use real weather observations as their starting point. Rather, they start with a very short-range prediction – usually just a few hours – from an earlier run of the model, which is then adjusted as necessary to fit in with the latest actual observations. This process is called initialization, and it is necessary to minimize uncertainties caused by the uneven distribution of observations over the Earth's surface and throughout the depth of the atmosphere, to eliminate spurious observations and errors in the communication of the data, and to obviate the effect of atmospheric gravity waves. Climate models are increasingly initialized in the same way, but the absence of observational data in the past over the oceans, deserts and ice-caps, as well as aloft in the atmosphere, means that our detailed knowledge of changes in the climate even in the very recent past is seriously limited. This is very important when climatologists are trying to fine-tune the model by hindcasting – that is, running it backwards to see how well it mirrors changes in our planet's climate over the last several decades or centuries.

So we can seen that there is a wide range of effects which limit the ability of a climate model to predict climatic changes in the future. Further, we cannot easily calculate how well the model is working by checking its predictions against reality – for obvious reasons – except by hindcasting. We should always bear in mind these imperfections and limitations when new model predictions are published.

Second-hand reports of the ponderings of our climatological experts which we might read in newspapers and magazines, or hear on radio and television programmes, suffer from a further series of problems. Climatologists are usually very careful in the language they use in their reports, using words like 'may' and 'could', and phrases like 'is predicted to' and 'within the limitations of the model can be expected to'. News reports frequently – to their authors' immense discredit – replace all these with 'will'. Original reports may refer to a predicted rise in temperature 'between 1.5 and 5.5 degrees Celsius', these figures representing the outer limits of a wide range of outcomes depending on differing initial assumptions of, say, greenhouse gas emissions. News reports often prefer instead to say 'up to 5.5 degrees Celsius', or worse still 'a temperature rise of 5.5 degrees Celsius'. By their very nature, the limiting values of a large number of possible outcomes are the least likely to occur. In this way, partly through ignorance and partly through the attractiveness of reporting extreme values to our often sensationalist news media, the general public are misled and scientists traduced.

Can we make a 100 year forecast?

Given what most of us understand a forecast to be, the answer to the question has to be no. The coarseness of the models, the uncertainty of their output, and the range of results dependent on initial assumptions all mean that, even at a worldwide level, published predictions can only be treated as an approximate guide to what might happen. National governments, usually driven by short-term self-interest, need more information at a much more local level to help determine their stance on this global conundrum. Some models attempt to predict regional variations in temperature and rainfall, but it is difficult to accept that these are yet much better than guesswork.

It is true that many aspects of the climate of certain areas on the Earth's surface are strongly dependent on temperature: for instance, a rise in surface water temperature over the tropical north Atlantic of 2 or 3°C (3.5 to 5.5°F) would probably result in a greater number of hurricanes in the Caribbean/USA sector, occurring over a longer portion of the year, affecting a larger geographical area, and some of the storms could well be more intense than any seen today. However, in most places the character of the climate is determined by the interaction of several factors. For example, a warmer world may well result in high pressure systems being more prevalent over the British Isles in summer, and at first glance one might expect that to result in lower summer rainfall, but at the same time the increasing temperature of the shallow seas around the islands would encourage greater humidity and an increased threat of widespread and torrential rain during thundery episodes.

The easiest trap a commentator on climate change can fall into is to try to simplify the climatologists' reports into a sort of forecast for the next 50 or 100 years. Thus we may be told that temperatures will climb progressively during the twenty-first century and by the year 2100 they will be up to 6°C (10°F) higher than they are now; at the same time winters will become wetter and summers will be drier (there's that word 'will' again). Alternatively the commentator may attempt to equate a regional climate at the end of the century with a different regional climate at the present time: for instance, the British public is frequently told that the climate of southern England will be like that of Bordeaux by 2050, and will resemble that of the western Mediterranean by 2100. Such indications are usually larded with positive allusions to the likely future success of British vineyards and beach holidays, perhaps with the aim of insinuating to a suggestible readership that future climate change might be a good thing.

It is easy to understand why this happens. The scientific publications reporting on the findings of climate prediction models are frequently as dry as dust, couched in very careful language, and littered with uncertainties, mathematical probabilities, and other caveats. In other words they are usually very difficult for the general reader to understand. The newspaper or television correspondent needs to grab the attention of his audience, and then to present a clear, comprehensible picture – even if it is wrong.

A different approach is called for here. Using the output of climate prediction models together with the expert opinion of climatologists and other environmental scientists, we are going to report the major climate-related events of a notional twenty-first century just as a newspaper might do. It is not in any way intended to be a forecast or prediction, merely an indication of the sort of events which might occur, and decisions which might have to be taken, during the next 100 years. The aim is to present a clear and comprehensible picture of a possible future – without pretending to predict the future.

Chapter 24: Press cuttings from the London *Daily Telegraph*, 2000 to 2100

London February 12 2008

Britain is again in the grip of ice and snow following the briefest of thaws last week. Over the weekend a north-easterly gale blowing directly from the Russian Arctic brought heavy snowfalls to eastern and central England, and yesterday the temperature remained below freezing over the entire British mainland for the first time in 20 years. Coldest spots were Aviemore and Braemar in the Scottish highlands, where the mercury sank to minus 16°C (3°F) early on Sunday morning.

A spokesman for the recently privatised WeatherUK, formerly the UK Meteorological Office, explained that this is already the coldest winter since that of 1978–9, and if the Arctic winds continue for another two or three weeks it may yet rival the historic winter of 1962–3 as the coldest of the last 200 years. The big freeze set in during the weekend before Christmas and tightened its grip during the New Year as a succession of heavy snowfalls swept southern England, Wales and the Midlands. The frost loosened its hold briefly around January 23 and again on February 5–6, but many parts of the country have now been continuously snow-covered since Boxing Day, a period of 50 consecutive days.

Climate experts are busy trying to reconcile this exceptional winter (and also last year's poor summer) with global warming. One senior scientist explained: 'An upward trend in global temperature does not mean that the temperature climbs steadily year-by-year in all parts of the world. Cold episodes will continue to occur in one part of the world or another, and this year it is our turn. Global warming does not preclude occasional severe winters in Britain. It is simply that they are becoming rarer as the years pass. The last really cold winter was in 1979, 29 years ago, but we had one 16 years before that, in 1963, and also 16 years before that, in 1947.'

Washington, December 31 2011

Climate experts at the National Oceanographic and Atmospheric Administration (NOAA) said that, globally, this year has been the warmest on record, beating the previous record which was established last year by a huge 0.4°C (0.7°F).

After it appeared that the global warming trend had slowed somewhat since the beginning of the present century, this sudden upward surge provides new ammunition to those nations who wish to rekindle the international discussion of climate change following the collapse of the Kyoto environmental summit a decade ago. However, the present US administration shows no inclination to engage in talks, despite pressure from the European Union. Nine of the ten warmest years since comparable records began in 1860 have occurred since 2000.

Scientists say the intense El Niño phenomenon in the tropical Pacific which began last April has contributed to the dramatic jump in global temperature this year.

Sydney, February 2, 2013

The New South Wales premier has declared a state of emergency in several districts in the Blue Mountains, bordering Sydney's western suburbs, brush fires rage out of control. Fires are also burning in two areas to the south of Canberra. Several homes have already gone up in flames, and an elderly couple died before they could be rescued. The emergency powers enable the state government to forcibly evacuate areas considered at risk.

This is the third successive summer to bring serious fires to south-eastern Australia, and follows the most intense and long-lasting drought in the nation's history. In Sydney itself rainfall during the last 24 months has been 40 per cent below normal. Most prolonged droughts in this part of Australia are linked to the El Niño phenomenon in the Pacific Ocean, and this year is no exception.

Los Angeles, March 4, 2013

Los Angelinos know what to expect when the TV weatherman tells them there's an El Niño brewing up. They may not understand what El Niño is, but they know it means stormy winters in southern California. This has been the second wild and windy winter in succession in this land famed around the world for its blue skies and sunshine, and rainfall totals for the season have been the highest since 1983 – another El Niño year. Homes in the Santa Monica mountains have been demolished by landslips, and mudslides have closed the Pacific Coast Highway for several weeks. At least the water companies are happy – the reservoirs that serve this huge urban sprawl are brim full.

Quito, April 13, 2013

Flash floods ravaged the Ecuadorean capital yesterday following torrential downpours which dumped almost 500 mm (20 in) of rain on the city in 48 hours. Shanty towns on the hill slopes above Quito have been destroyed, and the authorities say that over 500 people have died – either carried away by the floodwaters or buried under mudslides. The storms have also hit neighbouring Peru, where the death toll has reached 100. The succession of disastrous floods which have hit these two nations during the last two years have left both economies in tatters. Billions of dollars of emergency aid have been directed here by the United Nations Disaster Relief Agency since the beginning of 2012.

This latest catastrophe to hit this part of South America is directly linked to the current El Niño episode which is now entering its third year. Previously, no intense El Niño has ever lasted as long as this, although weak El Niño conditions persisted for four years in the early 1990s. Climatologists are now convinced that El Niño events have become longer, more frequent, and more intense, because of the global warming trend.

Bermuda, December 7, 2014

Hurricane warnings were lifted last night as Tropical Storm 'Wilfred' brushed past the island group and headed northeastwards into the open waters of the Atlantic Ocean. Sustained winds measured at Kindley Field airport reached 52 mph with peak gusts of 70–75 mph, and heavy surf pounded north-facing shores. Wilfrid may not have been a major storm, but it was certainly a record-breaker in other respects. It was the twenty-first named storm of the 2014 season in the Atlantic-Caribbean sector, and this was therefore the first time that the letter 'W' had been used to name a tropical disturbance here. The previous record was 19 named storms in 1995. It was also the latest in the season – in fact it is now more than a week after the official season's end. Forecasters predict that the remnants of Wilfrid will continue to travel north-eastwards, weakening further, before being absorbed by a middle-latitude Atlantic depression early next week.

London, July 14, 2029

The temperature soared above 100°F for the first time ever in Britain today as the heatwave intensified. At Heathrow airport, the hottest place in the country, the mercury touched 38.2°C (100.8°F), beating the previous record of 37.5°C (99.5°F) which was set at Mildenhall, Suffolk, in August 2011.

Scientists at the European Centre for Climate Change Observation (ECCO) say that, if the present warming trend continues, a temperature of 40°C (104°F) or more could occur in this country before the end of the century.

Weather forecasters expect the heatwave to break tomorrow as severe thunderstorms spread northwards from France.

Miami, September 8, 2033

Disaster struck the Cayman Islands on Wednesday when Hurricane 'Kate' scored a direct hit. The storm, already a Category Five hurricane when it caused extensive damage to the south coast of Jamaica on Tuesday, intensified further as it approached the island group. Remote-sensing meteorological equipment directed through the eye of the storm measured sustained winds of 180 mph, with gusts to 220 mph. It was therefore the most violent Caribbean hurricane on record – more intense even than Gilbert in 1988 or Larry in 2020. No lives were lost on the Caymans thanks to the efficient evacuation of the islands following the Double Red Precision Warning issued by the International Hurricane Center in Coral Gables, Florida. However, first reports indicate catastrophic damage to the infrastructure, with many buildings completely destroyed.

Melbourne, January 28, 2034
The Bureau of Meteorology confirmed this morning that a new Australian temperature record has been established. The mercury soared to 53.3°C (127.9°F) at Bourke in New South Wales yesterday, beating the previous record which was set 11 years ago at Marble Bar in Western Australia.

Ottawa, April 2, 2034
Canadian scientists confirmed that Hudson Bay failed to freeze over last winter for the first time on record. Extensive ice shelves formed around the bay's coastline from December onwards, but the consistently mild winter weather prevented the general freezing which normally lasts from late December to late March. The date of break-up of the ice in Hudson Bay has become progressively earlier during the last four decades as global temperatures have risen, and has become an important marker of our changing climate to the nations of North America. This year, though, indicates a worrying acceleration of the warming trend; climate experts had not expected a winter as free from ice until much later in the century.

Berlin, April 10, 2034
Heavy snow blanketed central Europe today as a vigorous depression moving north from the Mediterranean encountered a bitterly cold north-easterly blast drawing air directly from Siberia. Transport systems were disrupted in many parts of the EU from Warsaw and Budapest in the east to Lyon and Dijon in the west, while heavy downpours caused serious flooding on the French Riviera and in northern Italy.

Weather experts said that such freak wintry conditions used to occur quite frequently in April in many parts of Europe, and occasionally even in May, but have become much rarer in recent decades. They emphasized that this present spell, although unusual by modern standards, did not represent any slowing of the global warming process.

San Francisco, June 21, 2034

An unprecedented heatwave has gripped northern California for over a week, and several long-standing temperature records have been smashed out of sight. In San Francisco the mercury climbed to 40.6°C (105°F) on Friday, and to 41.7°C (107.1°F) yesterday, eclipsing the previous record which was set back in 2001. In Sacramento a high of 48.3°C (119°F) was logged, beating a record established in 2005. Hundreds of people in both cities were hospitalized, suffering from sunstroke.

Rio de Janeiro, March 19, 2041

A hurricane crossed the Brazilian coast last night between the towns of São Mateus and Linhares in Espirito Santo province. Winds averaging 75 mph caused considerable damage, heavy seas pounded the coastline, and the coastal and main north-south highway was blocked by floodwaters and mudslides.

Forecasters said that the hurricane would weaken rapidly as it moved inland. The tropical storm, first observed to the west of Ascension Island last Wednesday, was the first ever recorded south of the equator in the Atlantic Ocean. Experts say that the water temperatures in the south Atlantic have not been high enough to support the development of a hurricane until now.

Climatologists predicted as long ago as the 1980s that hurricanes would eventually form in this region if global warming continued – the only surprise is that it took so long. Brazil is the country most at risk from hurricane strikes in tropical regions of the south Atlantic, and this latest development will give added impetus to the establishment of a hurricane warning centre here in Rio.

Jerusalem, January 22, 2042

The joint Israeli/Palestinian capital was covered in a layer of snow for the first time in 22 years on Friday as storms blocked traffic in many parts of the two nations. The storms swept down from the north, having already dumped substantial amounts of snow over parts of Syria and Lebanon, causing chaos on the roads and resulting in the cancellation of several regional flights. Electricity supplies to the hillier parts of Jerusalem were interrupted overnight. Torrential rain at lower levels led to flooding, notably in Tel Aviv. Climatologists said that heavy snow used to be a significant winter hazard in Jerusalem, occurring once every two or three years during the twentieth century, but this was the first major snowstorm in the city since March 2020.

Madrid, January 26, 2042

Ten people have died and hundreds hospitalized suffering from respiratory problems after a severe duststorm swept the region of Andalucia in southern Spain and the Algarve in the south of Portugal, EU health officials said on Tuesday. Thousands of people across Spain and Portugal have experienced breathing problems, sore eyes, and skin irritation caused by the fierce southerly wind, known locally as a *leveche*, which is laden with dust carried from the heart of the Sahara desert. The dust also cut visibility, causing problems for road, sea and air traffic. Such duststorms have become more frequent in many parts of southern Europe in the last few decades.

Buffalo, February 28, 2042

It's official: Buffalo is still the snowiest city in North America. A report in the *Buffalo News* yesterday quoted figures from the National Weather Service which show that the city still experiences heavier and more frequent snowstorms than any other 'million city' in the continent. However, recent fluctuations in the climate have brought changes to the pattern of snowfall in the city.

Always at risk from so-called lake-effect snowstorms, the relatively warm waters of Lake Erie used to supply the moisture for regular early winter storms in Buffalo. In 2001 a five-day snowfall dumped over two metres (seven feet) of snow in the district. Later in the season the Lake used to freeze over, cutting off the supply of warmth and moisture and reducing the frequency of snow.

Nowadays it is rarely cold enough for snow in December, and Lake Erie never freezes over completely, so Buffalo's big snows come in January, February and early March rather than December.

Milan, March 3, 2042

Two months of blue skies in northern Italy have pushed pollution concentrations to a critical level, prompting a ban on private vehicles in the financial centre of Milan. Drought also threatens the region's farmers. In some areas no rain has fallen for 100 days. Weather experts say that Milan and Turin have both recorded their driest winter on record. Newspapers on Tuesday showed pictures of dried-out Venetian canals with boats languishing in the mud. A regional health spokesman said that measures to reduce winter pollution in northern Italy had been broadly successful during the last half century, but conditions this winter had been extreme, leading to accumulations of carbon monoxide and nitrogen oxides not seen since the 1990s.

Washington, June 24, 2047

Antarctica continues to confound climate experts. Huge chunks of ice have been dropping off the West Antarctic ice sheet for the last 60 years, but a new report from the United Nations agency, the World Climate Change Organization, confirms that the ice over the bulk of the Antarctic continent continues to grow thicker.

The weather in south polar regions is now monitored in great detail, and although the temperature has risen by an average of 4°C (7°F) on the Antarctic peninsula since the beginning of the century, no significant warming has yet occurred in the continental interior. At the same time snowfall has become more frequent.

The WCCO report describes how satellite-based radar systems can now measure changes in elevation to an accuracy of a millimetre. These measurements show that the quantity of ice over the Antarctic continent has increased by two per cent in the last 25 years. Experts believe that the strengthening of the westerlies over the Southern Ocean has created ever deeper depressions around the fringes of Antarctica, but this has effectively isolated the bulk of the continent from the warming trend which is affecting most of the rest of the planet.

Geneva, January 2, 2050

Figures collated by the World Meteorological Organization show that global warming continued during the 2040s at approximately the same rate as in earlier decades. Average global temperature in the last decade stood at 16.8°C (62.2°F) compared with 14.8°C (58.6°F) in the 1990s. This means that the rise in temperature worldwide amounts to roughly 0.4°C (0.7°F) per decade over the last 60 years.

The twentieth annual report of the World Climate Change Organization announced the latest prediction from the world's leading climate experts. They continue to call for a slight acceleration in the warming trend during the remainder of this century, and forecast global temperature to be between 19.0 and 19.5°C (66.2 to 67.1°F) by the year 2100.

The WCCO claims some success in recent years at stabilizing the emission of greenhouse gases. Industrial processes in the rapidly expanding economies of south and east Asia have become much cleaner during the last 10–15 years thanks to the Bangkok Agreement of 2035 which initiated the sharing of emission-free technologies by the world's most advanced countries.

Moscow, January 25, 2050

Muscovites have not needed their fur hats and coats this month as temperatures have soared day after day to record levels for January. With afternoon readings exceeding the freezing-point every day but one so far this month, residents of the Russian capital have squelched through puddles of rain-water instead of trudging over piles of snow. Many have even got used to carrying umbrellas.

The thermometer yesterday showed 7°C (45°F), equalling the reading of the previous two days, a figure which has never been approached before in January in 155 years of records in Moscow. Snow has fallen just three times since Christmas.

Santiago, June 17, 2060

Gales battered the southern part of South America over the weekend, killing at least 300 people as the wind ripped roofs off houses, disrupted traffic on land, at sea and in the air, and left millions of homes without power. Wind gusts of up to 135 mph tore through southern and central Chile, before crossing the Andes and ravaging large areas of Argentina and Uruguay.

A ferry, unable to enter the harbour at Montevideo, the Uruguayan capital, because of rough seas, capsized with the loss of over 200 lives.

Heavy rains also swept parts of Chile leading to flash floods and mudslides, and the main route from Santiago to the southern provinces of Aisen and Magellanes was blocked. Hundreds of people in the city of Concepción were evacuated from their homes as flood waters tore through low-lying suburbs.

Nice, September 19, 2062

Rain doused most of France's 'Black Summer' forest fires, and relieved firefighters said that the crisis was nearly over. No lives had been lost, but about 500 homes and 650 square kilometres (255 square miles) of forest and farmland were destroyed by the fires, many of them burning since the middle of August. The insurance bill is expected to exceed 10 billion euros.

Some 25 to 50 millimetres (one to two inches) of rain fell across central and south-eastern France overnight on Thursday, and was heaviest in the areas where wild fires still burned, so that firefighters were finally able to bring them under control. These rains also caused minor flooding, cut power supplies to about 50,000 homes, and blocked several roads with mud and water.

Hailstones the size of tennis balls shattered windows, windscreens, and roof tiles in the city of Avignon.

Cambridge, July 23, 2065

The death has been announced of the prominent English global warming sceptic, Dr Roy Fitzrobert, at the age of 60. It is believed that he took his own life. For most of his professional life he found himself in a small and diminishing minority of scientists who believed that the greater part of the warming trend in the last 80 years was a consequence of natural climate change, but he revelled in the notoriety, and his position as the Director of the Darwin Institute in Cambridge gave him a platform from which he was able to publicize his views for over 30 years. Friends say that he became increasingly depressed following his retirement earlier this year. He died on the hottest day of the century in England, with the temperature touching 39.3°C (102.7°F) in Gloucester.

Dacca, November 17, 2070

It is feared that over a million people died in the cyclone and tidal wave which brought death and disaster to Bangladesh last week. Only now, as the waters recede, is it possible to see fully the devastating effect the storm has had. The tidal wave – or storm surge – was six metres high, the same height as a two-storey building. It has changed the map of the delta region which encompasses much of the regions of Khulna, Patuakhali, Barisal and Noakhal, sweeping away some islands and making others. Whole communities have been destroyed and all their people and livestock killed.

One week ago, Hussain Ali, a 35-year-old local administrator, had almost 70 family members all living in newly-constructed houses in the village of Medua. Today only 12 of them are alive. 'We were all sleeping when the tidal wave hit,' he said. 'I grabbed hold of a palm tree and held on until the waters subsided around daybreak. I think I was there for about six hours.' The United Nations Disaster Agency went into action immediately, and a huge effort to rescue people and deliver aid is now under way. Shuttles flew into Dacca today carrying vehicles, food and medical supplies, but it will be some time before this aid can be transported to the worst hit areas. All roads in the delta region have been washed away, and the currents are too swift and unpredictable in some of the channels – now uncharted – for relief boats to get through.

This cyclone struck 100 years almost to the day after, and in the very same place as, the great disaster of 1970. It had been monitored closely for several days and was predicted to make landfall in the relatively sparsely populated coastline of north-western Burma. But a last-minute change of direction allied to a sudden intensification of the storm sent it into the populous delta area. The region was on cyclone alert, but no general evacuation had been ordered. Hundreds of thousands of people had decided to spend the night in municipal cyclone shelters as a precaution; the death toll would have been even higher if they had not.

Los Angeles, August 31, 2074

The USA is closing its border with Mexico with effect from midnight tonight. The torrent of migrants heading north from the growing deserts of Central America has reached a level –

now over 5000 a day – which has forced the hand of the American government.

Mexico, a nation of 180 million people, is finding it increasingly difficult to feed its teeming population. Rapid industrialization during the early decades of the century provided the wherewithal to import a growing proportion of the country's food requirements, but the pressures of climate change in many parts of the world has sent the price of many staples soaring through the roof. The gap between the gross national products of Mexico and the USA actually narrowed slightly between 2040 and 2060, but it has widened again in the last decade.

Mexico has suffered more than most from the ravages of the new climatic régime. There is an increased incidence of hurricane strikes on both the Caribbean and Pacific coastlines, temperatures regularly exceed 50°C (122°F) in lowland districts, and declining rainfall in the country's northern states of Sonora, Chihuahua, Sinaloa, and Durango has led to serious water supply problems.

London, April 13, 2081

Oceanographers at Southampton University in southern England have observed disturbing fluctuations in the large-scale ocean currents of the north Atlantic Ocean during the last five years. Measurements of the speed of flow of the North Atlantic Drift, an extension of the Gulf Stream, which is largely responsible for the mild and rainy winter climate of north-west Europe, have revealed a dramatic slowing during the last year.

Scientists have long speculated that the present warming trend might be fragile in certain parts of the world, including western Europe, because the world's changing climate might cause sudden and unpredictable changes in the pattern of ocean currents.

The Southampton group, one of several studying these currents in the Atlantic Ocean, are convinced that the recent fluctuations in the North Atlantic Drift are a response to this century's planet-wide warming. Computer models are still unable to say for certain what will happen in future decades, but Professor Steve Lincoln, the group's leader, warns that dramatic changes in Europe's climate could occur. 'What is fairly clear is that if the ocean circulation patterns which provide winter warmth around the northern and western sides of the north Atlantic were to slow or stop, the consequences could be very serious,' Lincoln said. 'This could also happen much quicker than people appreciate. We have to be aware of the risks.'

The North Atlantic Drift has slowed before in historic times, notably during the sixteenth and seventeenth centuries, and it probably stopped completely on a number of occasions in Earth's distant past. It appears to be particularly sensitive to increases in temperature and to influxes of fresh water. 'The same very cold, very salty water that sinks in the far north of the Atlantic Ocean simply won't sink if it's just a few degrees warmer or a little bit less salty,' Lincoln added. 'Our research shows that these changes have not happened gradually over a period of hundreds or thousands of years, but suddenly, within 20 or 30 years. The system does not respond in a linear manner. It's as if a touch on the brakes switches the engine off. The circulation pattern slows down faster and faster and may eventually stop altogether.'

The paradox, speculated upon as long ago as the 1980s, is that the same greenhouse effect which is making the Earth warmer overall, could have the opposite effect on much of Europe by slowing down – or even shutting off – the warm ocean current on which it depends.

New Hampshire, March 4, 2082

An American institution died this month. The rising temperatures and diminishing snowfall of recent decades in New England have finally seen off maple syrup production in the

USA. Climatic conditions are the major factor in making maple syrup, said Karl Andersen, director of the New Hampshire Department of Agriculture. Nights with temperatures of −5°C (23°F) or below and daytime temperatures around freezing are needed for peak production. Trees are tapped and most maple sap is collected during February and March. Yields have been decreasing for several decades, and several recent winters have been simply too warm for commercial quantities of the syrup to develop. All American supplies of maple syrup will now be imported from Canada.

Montana, March 21, 2082
Facing a tenth successive year of drought, Montana has returned to the Dust Bowl Days of the 2020s and 2030s, the senior state meteorologist told the Montana Drought Advisory Committee. Rivers and reservoirs are at record low levels, mountain snowpack is vanishing into thin air, and water supplies for domestic and commercial use are being piped in from Washington and Oregon. The drought is just as bad in the neighbouring states of Wyoming and the Dakotas. Throughout the region topsoil is being blown away on strong winter winds, and the frequency of duststorms at the meteorological station at Bismarck, North Dakota, has increased tenfold in the last 50 years.

Paris, February 11, 2091
Europe is enduring its coldest winter for decades. Biting easterly winds have swept across the continent from northern Russia, and severe snowstorms have hit France, Britain and Germany. Paris has been continuously snow-covered since December 27, and London for almost as long.

Weather experts say that the return of wintry conditions has been caused by a huge 'blocking' high pressure system which developed over Scandinavia around Christmas, and has not budged since. As a result, Atlantic depressions have been diverted well south of their normal track between Scotland and Iceland; instead they have travelled south-eastwards into the Mediterranean Basin. Forecasters see no end in sight to this remarkable cold spell.

Ocean scientists say that sea-surface temperatures between Scotland and Iceland have dropped 5°C in the last ten years, although globally-averaged temperatures are as high as ever. At the same time ocean currents such as the Gulf Stream and the North Atlantic Drift are flowing ever more slowly. It is now widely accepted that global warming is having a serious knock-on effect on currents in the north Atlantic ocean, which in turn is predicted to create a sharp localized cooling of the climate over western Europe.

Edinburgh, August 31, 2091

Scotland has experienced its coldest summer in over a century. Provisional statistics indicate that in Edinburgh the mean temperature for June, July and August has been the lowest since 1965. Sunshine has also been in short supply, but rainfall has been well below the long-term average. With frequent north-easterly winds grey drizzly weather has been the rule, and heavy downpours have been notable for their absence. Mountain walkers have noted several snowbed survivals in the Cairngorms this summer for the first time since 2008.

The cool summer has not been confined to Scotland. Other parts of the British Isles, northern France and the Benelux countries have also reported their lowest temperatures for several decades, and in Amsterdam the temperature failed to reach 30°C (86°F) for the first time since 2022. However, eastern Europe and northern Scandinavia have experienced a very hot summer with southerly and south-easterly winds frequently carrying air of tropical origin to regions inside the Arctic Circle.

Tokyo, January 1, 2100

Scientists working at the New Hydrogen Energy Laboratory in Sapporo, northern Japan, will today demonstrate experiments which create energy from the fusion of deuterium nuclei – so-called 'cold fusion'. It was in 1989 that American and British researchers at the University of Utah first claimed to have carried out successful cold fusion experiments. For a few heady weeks the world thought that the prospect of endless energy, at negligible cost and with hardly any pollution, was just around the corner. However, other scientists were unable to replicate the experiments, and the work of Stanley Pons and Martin Fleischmann was soon derided by the bulk of the scientific community. Some even shouted 'fraud'.

For many years work continued in isolated localities, notably in Japan and Italy, but it was poorly funded and little more was heard of cold fusion until the 2080s. During the last two years the cold fusion team at Sapporo has published several papers in major on-line journals, and has scrupulously avoided the mistakes of Pons and Fleischmann by making no public announcements until their work had been closely examined by other – often highly sceptical – physicists and chemists. Last year, teams of researchers from Europe and America were invited to Japan to ensure they were able to reproduce the results of the NHEL scientists.

Much progress has been made in the last 25 years to neutralize industrial emissions, but this remains a very expensive option. If energy from cold fusion becomes commercially available early in the new century, it will help the nations of the world to cut the amount of carbon dioxide, methane, and other greenhouse gases in the atmosphere at a much faster rate and at a much lower cost. It will still take a long time for the world's climate system to stabilize, however, and we shall no doubt have to endure further major natural disasters in the years to come. But with the new century comes new hope.

Chapter 25: Climate change and water supply

Pure fantasy this may be, but it is important to get a feel of the sort of pressure that accelerating climate change could exert on the twenty-first-century world. A rising global temperature would certainly present problems, but it is the climatic and other environmental changes triggered by this warming process which might spell disaster for a large number of nations around the world.

Let us highlight the most obvious of these: the general poleward movement of the world's climatic zones. Add to this the gross economic and social inequalities across the planet which are not going to disappear in a hurry during the coming century, political differences which may or may not be reconcilable, and a continuing growth in population, especially in the economically poorer regions, and you have a perfect recipe for widespread social unrest.

A broadscale northward shift of the climatic zones in the northern hemisphere (southward in the southern hemisphere) will change the distribution of rainfall over much of the planet. Rain is easily the most important of the various meteorological elements for human survival. Our supply of water for drinking and bathing depends on rain, so does water for industrial and leisure activities, and for growing crops and raising livestock. In areas where rainfall is not sufficient for crops to be grown or animals to be reared, water may be imported by irrigation; for instance, it can be brought from wet mountain ranges to dry plains – as long as the mountain rains do not fail.

Even in places where rain normally falls regularly and copiously, in Britain or Ireland for example, or along the eastern seaboard of the USA, a dry summer brings problems. Springs and wells fail, reservoirs dry out, the usually green countryside turns brown, and the supply of water to every home – taken for granted for most of the time – may be rationed. In lands where seasonal rains are the norm, such as

India and China, a delay or a false start to the rainy season will seriously affect the amount and quality of the food grown. A failure of the rains may lead to famine.

An excess of rain is also unwelcome, washing away crops, damaging property, blocking transport arteries, and causing disruption in urban areas. Flooded rivers burst their banks, drowning livestock and making farmland unusable for months. In wealthy countries such events are rarely catastrophic, but in those where farmers can produce little more than they consume, and where stocks of food for the general populace may be insufficient, the consequence will again be famine.

Shift the world's rainfall distribution around, and you get large areas with a serious water deficiency and equally large areas with a serious excess. A northward extension of the sub-tropical high pressure systems in the northern hemisphere could well result in the Sahara desert growing into the fertile Mediterranean lowlands of Morocco and Algeria, and then into the already dry lands of southern and eastern Spain. One could envisage a diminishing of winter rains in the Near and Middle East leading to harvest failure in Israel, Lebanon and Syria, while the steppes of Russia and the Central Asian republics turn to sand.

The deserts of north-west Mexico and the south-western USA would also be expected to expand, and the exhausted soils of America's bread-basket states could simply blow away after a succession of dry years. In the southern hemisphere the Australian Desert might engulf some of the pastures of south-western Queensland, western New South Wales, and northern Victoria, the Namib desert could extend southwards into South Africa, and the Atacama would be expected to grow towards the Chilean capital, Santiago. And where might there be more rain? Probably in north-west Europe, western Canada, south-eastern USA, India, China and south-east Asia.

Those regions which become drier will have to deal with a growing demand for water and a diminished ability to supply it. The situation will be exacerbated if the population is growing. Some commentators have already postulated that the regional wars of the early twenty-first century will be water wars. Where rivers which flow across several national boundaries provide the only reliable year-long source of water, any scheme to control the flow of water in one country could be regarded as an act of aggression by the next country downstream. Already we have seen tension between Turkey on the one hand and Syria and Iraq on the other, as a consequence of Turkish plans to control the flow of the headwaters of the Tigris and Euphrates rivers. Israel is reluctant to relinquish the occupied Golan Heights because it would yield some of the streams which supply water to the river Jordan to Syria.

Where the economic and social stress of diminishing water supplies becomes too great, a movement of population will ensue. If food supply also fails, mass migrations could be expected, on a scale which would make the economic refugee problem of the early years of the twenty-first century look like a vicar's tea-party. One could envisage millions trekking north from Central America to the USA, and from north Africa and the Near East to southern Europe.

The pressure of these climatic refugees on the wealthy nations of North America and Europe would come at a time when growing climatic stress might be expected in these regions too. Even in the countries of the European Union, and in the USA and Canada, it is likely that the regular and reliable supply of water will become a recurring political issue, rather like transport policy at the turn of the century. Metering would become universal, costs would rise steeply, and occasional bouts of rationing would become more frequent. Add to this the problems presented by rising temperatures, rising sea-level, and the possibility of dramatic fluctuations in climate in response to changes in ocean currents, and it can be seen that, if the predicted changes in global weather patterns are even half right, climate will worm its way into almost every aspect of social and political life.

Chapter 26: Future imperfect

One of the well-known paradoxes of weather forecasting is that it is often easier to produce a reasonably accurate general forecast for seven days ahead than it is to compile a detailed forecast for the next 24 hours. The same superficial contradiction can be applied to climate prediction too. It may be impossible to say with any precision where global climate will go from year to year – or even from decade to decade – during the twenty-first century, but we can probably indicate the broad trend during the coming 100,000 years. There are several big assumptions here, though. To what extent will unintentional anthropogenic climate change make a significant contribution to the world's atmospheric environment, and for how long? Will human beings ever be able to exercise control over the Earth's atmosphere, that is, intentional anthropogenic climate change? If so, when, and will they wish to control it? Will there be any catastrophic event such as a major asteroid impact during this period?

From our standpoint we can only reasonably work on the basis that unintentional anthropogenic influences are reversed, and that the other events do not come to pass. In other words we simply consider those climatic controls, mainly cyclic, that we know have dominated our planet's climate in the past. It is certainly possible to extrapolate these factors for at least one more glacial/interglacial cycle into the future and therefore to predict how the atmosphere responds to them. We can therefore produce a broad-brush sequence of climatic changes for the planet as a whole and for the individual continents as well. The results of this sort of research present us with a background of natural climate change against which fluctuations induced by human activity can be viewed; they also give us some idea of the environmental problems which may be faced by our descendants in the distant future.

In an earlier chapter we investigated in detail the story of the world's climate since the end of the last glacial period some 11 millennia ago, we outlined the major factors involved in the generation of these climatic fluctuations, and we saw how important the various astronomical cycles are in this regard. By their very cyclicity these astronomical factors are inherently predictable, perhaps for millions of years into the future. Now we can see that all that painstaking research building a picture of past climates and relating it to those astronomical periodicities can help us to create a picture of future climates too. At its simplest level, the daily revolution of the Earth on its axis shows us that the temperature is normally higher at 2 pm than it is at 2 am, and we can therefore make a prediction that over the next 1000 days it will be warmer at 2 pm compared with 2 am on a large majority of occasions. Similarly, the annual orbital cycle of the Earth around the Sun reveals that, in New York and London and Tokyo, July is warmer than January in every year, so once again we can make a forecast with a high probability of success that July will be warmer than January in every one of the next 1000 years. In the same way we now know that lengthy glacial periods have alternated with relatively brief interglacial episodes on a cycle of 90,000 to 100,000 years, and this allows us to predict that the present interglacial will before too long give way to renewed glacial conditions which will probably last for the best part of 100 millennia. In this way, then, does the past tell us what the future holds.

More detail than this is impossible to predict. We may be able to say that tomorrow will be warmer than tonight by assessing the effect of the Earth's rotation, but it does not tell us whether tomorrow will be warmer than yesterday. And it certainly gives us no inkling what the remaining details of the weather will be like. These are under the control of physical forces within the atmosphere itself, which manifest themselves as depressions, fronts and anticyclones on the meteorologists' weather charts. And so it is with long-term climatic prediction.

We may be able to predict the return of glacial conditions to middle and high latitudes of our planet, but the year-by-year, decade-by-decade, and century-by-century detail is in the gift of such events as volcanic eruptions, fluctuations in the energy emitted by the sun (usually called solar variability), changes in the pattern of ocean currents, and asteroid/comet impacts. From past experience it is unlikely that other Earth-related factors such as continental drift or mountain building will come into play during a period as 'short' as 100,000 years.

The first seriously scientific theory linking astronomical and glacial cycles was propounded by the English astronomer James Croll in 1864, although there were no useful data concerning the timing of glacial episodes at the time, so he was unable to verify his theory. A Serbian astronomer, Milutin Milankovitch, developed the theory in a succession of published papers between 1920 and 1941, and nowadays it generally goes by the name of the Milankovitch Hypothesis, which is a trifle unfair on Croll. By the 1940s we knew fairly accurately when the last glacial episode had ended, and also when the greatest extent of ice had occurred, but data about earlier glaciations were still qualitative rather than quantitative. Thus Milankovitch, like Croll before him, could not check his theory against reality. In the 1950s and 1960s, research workers began to tie down the chronology of the several glacial episodes during the Quaternary period – that is, during the last one million years – and the facts appeared not to support the theory. Thus the Milankovitch Hypothesis fell out of favour.

Rather more recently several palaeoclimatologists, including J.D. Hays, J. Imbrie, and A. Berger, have revisited the problem. They have been able to quantify feedback effects resulting from oceanic circulation patterns, in particular variations in temperature and salinity in the North Atlantic ocean gyre which appear to be especially important in controlling changes in northern hemisphere climate. Other feedback effects involve changes in the albedo (that is, the way different land and sea surfaces reflect solar radiation back into space; smooth, light-coloured surfaces reflect more energy than do rough, dark-coloured ones), changes in sea-level, changes in carbon dioxide concentrations in the atmosphere, changes in the Earth's biomass, and changes in ice thickness and extent in the Arctic and Antarctic regions. Most recently, André Berger, a professor at the Catholic University of Louvain-la-Neuve in Belgium, has produced a climate model based on the precession, orbital, and obliquity cycles, but which also takes into account the most important feedback effects. Although still quite crude in its parametrization of the Earth's surface and atmosphere, this model is able to explain very nearly all of the main fluctuations in our planet's ice volume during the last two glacial-interglacial cycles, and is notably successful in mirroring the sudden rapid warming which occurred at the end of both of the glacial phases.

Berger has also run the model forward, providing the first tantalizing glimpses of probable global climate variations in future centuries, assuming no prolonged anthropogenic interference. The results show how continental ice volume is predicted to change, and we may make a direct inference of sea-level changes from this. More detailed predictions of climatic characteristics at different places on the Earth's surface are, as yet, still qualitative, but we can probably make a fairly good stab at describing them by reference to the sequence of events during the last glacial-interglacial period.

At the present time the world's ice volume is approximately three million cubic kilometres, which is, with the exception of the period of the Climatic Optimum around 5000–3000 BC, the lowest it has been since the early part of the last interglacial phase, some 125,000 years ago. Berger's model indicates that the Climatic Optimum was indeed the peak of the present interglacial period, and that since then world climates have been edging slowly and erratically downhill. This process should continue for the next 4000 years, followed by a temporary stabilization; but the advance of ice and the associated climatic deterioration is then predicted to accelerate to reach a low point between 20,000 and 30,000 years from now. A slight improvement between AD 30,000 and 40,000 will be followed by a prolonged and rapid decline to a period of maximum glaciation between AD 55,000 and 60,000; the model suggests that this future glacial phase is likely to be of similar, or marginally reduced, severity compared with the last glacial maximum some 20,000 years ago. The warming after AD 60,000 is not predicted to be as large or as rapid as that at the beginning of the present interglacial period. Indeed, a further deterioration is indicated around 100,000 years from now, before the global climate finally emerges into the next true interglacial phase some 115,000 years into the future.

Years after the present	Continental Ice Volume (millions of cubic kilometres)	Sea-level (metres below present)
0	3	0
10,000	8	10
20,000	25	45
30,000	20	35
40,000	20	35
50,000	40	75
60,000	45	85
70,000	15	25
80,000	20	35
90,000	20	35
100,000	35	65
110,000	40	75
120,000	<5	0

We can translate these results into a descriptive account of the climate likely to be found in different parts of the world, although once again we should emphasize that the assumption

– an unrealistic assumption in the view of a majority of climatologists – is being made that climate change induced by human activity will not be a significant factor in the development of future climates. On that basis, the present reasonably benign environment would probably fade away very quickly, probably in the next thousand years, as the Arctic and Antarctic ice-caps grow, and the climate of middle and high latitudes in both hemispheres reverts to a state not seen since the Younger Dryas, some 10,000 years before the present day. Indeed, J. and K. P. Imbrie wrote in 1979: '… the present interglacial is already on its last legs, tottering along at the advanced age of 10,000 years.'

Thus by AD 3000, perhaps as early as AD 2500, we find the volume of land-based Arctic ice doubled compared with today. This means that the existing ice-caps on Greenland, the Svalbard archipelago, and some the northern Canadian islands grow thicker, and the ice advances across the coastal fringes which are at present ice-free in summer. Extensive permanent ice returns to the Cherskiy and Verkhoyansk Ranges and the Taimir Peninsula in Siberia, to Brooks Range in Alaska and the Mackenzie Mountains in north-western Canada, and probably also to the mountains of northern Scandinavia. Iceland once again becomes ice-bound and, along with the Greenland coast, probably uninhabitable under the social structures and technological knowledge of today. Glaciers would develop extensively over the Rocky Mountains as far south as Colorado and over the Coast Range of British Columbia and the Cascades in Washington and Oregon, over the high ground of Labrador and Newfoundland, in the Scottish highlands, and over the mountains of southern Norway, the Alps, and the northern Urals. Glaciers would also grow in the southern Andes, the Australian Alps, and the Southern Alps in New Zealand.

The deterioration in the heavily populated middle latitudes would present extraordinary social and economic stresses. Around the coasts of north-western Europe we can postulate a drop in sea-level of five or six metres, and also a drop in sea-surface temperature of at least 5°C compared with AD 2000. The change in sea-level may not sound very great, but around the British Isles it would be sufficient to reveal hundreds of square kilometres of new land in what are now the Thames Estuary, the Wash, Morecambe Bay, and Solway Firth, with a new island forming in the Straits of Dover where the Goodwin Sands are. The Netherlands, Germany, Denmark, and countries bordering the Baltic Sea would also find themselves growing substantially larger. The depression track across the Atlantic and northern Europe would be further south than it is now, resulting in a very changeable and sometimes stormy climate at all seasons, and the lower sea-surface temperatures would result in a substantial drop in air temperature at all seasons too. The inference is that, in central England, average January temperature would range between −4°C (25°F) at night and 2°C (36°F) by day, in July between 7°C (45°F) at night and 16°C (61°F) by day, while mean annual rainfall would be 25 to 50 per cent higher than at present. A typical winter would bring several heavy snowfalls, and the ground would be snow-covered for two months, while severe spring frosts and occasional snow showers would be quite common in May and June, and no month would be entirely frost free. Summers would be cool, cloudy, and often windy and rainy; although rare anticyclones would still permit occasional mainly short-lived spells of hot sunny weather, these would be much more uncommon than they are now. Gone are the twenty-first-century vineyards; gone, too, are the rolling fields of wheat and barley. If Britain still needs its agricultural base 1000 years hence, oats, potatoes, and pasture seem to be the limit imposed by the new climate. In north-west Scotland January temperatures would be similar to those in central England, but in July they would fluctuate between 5°C (41°F) at night and 11°C (52°F) during the afternoons, and the climate of Stornoway and Lerwick would be similar to that found along the north coast of Iceland during the present day.

A similar deterioration would be found throughout northern and central Europe; the Baltic Sea would be completely frozen over in most winters, while many ports bordering the North Sea such as Hamburg and Rotterdam would frequently be ice-bound. Cereal crops would grow best in lands bordering the Mediterranean, where the southward-shift of the mid-latitude westerlies would result in a longer rainy season, while summer drought would become confined to eastern Spain, and the eastern part of the Mediterranean Basin. The Mediterranean coastal lowlands of northern Africa could well once again become the bread-basket of the Old World as it was in early Roman times, with plentiful autumn and winter rains, and somewhat cooler summers. However, the deserts of subtropical latitudes – the Sahara, Arabia, and south-western Asia – would probably become even more arid than they are today.

In North America, changes would not be quite as dramatic as in Europe, except along the Pacific seaboard where the cooler oceanic waters would drag sharply downwards the temperature of places like Vancouver, Seattle, Portland, and San Francisco, while Oregon and northern California would probably catch appreciably more summer rain than they do at the present time. Inland in Canada and the USA, and also along the Atlantic seaboard, temperatures would probably drop 2 or 3°C (3.5 to 5.5°F) throughout the year, while increasing aridity in the Great Plains and the Mid-West would radically alter the agricultural capacity of the region. There would be little change in the frequency or severity of hurricanes or tornadoes.

In the southern hemisphere one would expect a more limited shift of the climatic zones. The growth of the Antarctic ice-cap would be largely confined to an increase in thickness; geographical expansion is limited by the rough waters of the Southern Ocean. Nevertheless, there would be a significant cooling of, say, a couple of degrees in the southern parts of South America, southern Africa and Australasia, and winter snows would become appreciably more common in Patagonia, on the High Veld in South Africa, in Tasmania and Victoria, and in New Zealand's South Island. A decline in rainfall in sub-tropical latitudes could well lead to some desertification around the fringes of existing deserts, especially in Australia, by this time.

Let us advance to the predicted peak of the next glacial period, some 58,000 years from now. The world, if it still exists, is a very different place, able to sustain life as we know it only between latitudes 50°N and 50°S. The Arctic ice-cap has expanded to 15 to 20 times its present volume, and permanent ice covers almost as much of the northern hemisphere as it did during the Wisconsin-Weichsel glaciation. A reasonable estimate of the ice-edge at its maximum extent would be across northern Russia, Belarus, Poland, and northern Germany, across the central North sea to northern England, Wales, and central Ireland, and in North America from southern New England to Lake Michigan, Montana, and Vancouver Island. Sea-level has dropped by as much as 90 metres (300 feet) from its present height, and the ocean waters have retreated from most of the continental shelves. The British Isles are again linked to the European continent, Siberia to Alaska, Sri Lanka to India, New Guinea to Australia, Japan to eastern Siberia, and several of the Indonesian islands to south-east Asia. Countries completely or nearly completely buried under ice include Canada, Greenland, Iceland, Scotland, Norway, Sweden, Finland, Denmark, Lithuania, Latvia and Estonia. The climatic zones which in AD 2000 lay between the equator and about 70°N/S are now squeezed into just 50 degrees of latitude in each hemisphere, and the winter storm track across the Pacific Ocean is typically around latitude 50–55°N, and if anything slightly further south than that across the Atlantic Ocean.

In Central England (or, with slight variations, in New York or Seattle or Paris or Vienna) the mean mid-winter temperature ranges from –35°C (–31°F) at night to –25°C (–13°F) during the daytime, while midsummer readings vary between 0°C (32°F) at night and 13°C (55°F) by day. From October until May the weather is often clear and dry, with many sunny days but also a relentless, bone-chilling east wind. Fierce blizzards intervene at intervals, maybe two or three times per month, but amounts of precipitation are relatively small. The brief summer season, lasting just two or three months, is an unpleasant time of the year; the ground, frozen solid until early June, becomes waterlogged with vast stretches of meltwater, and the weather is frequently damp and grey with a chill wind. Occasional incursions of southerly winds bring air from the Mediterranean region, and the temperature climbs dramatically, perhaps to 25–30°C (77–86°F) for a few days; but these brief heatwaves also bring an invasion of mosquitoes and other flying bugs. The first snows of late September followed by the big freeze-up in October come almost as a welcome relief from the all-pervading damp of the summer.

The most habitable regions in the northern hemisphere, as far as the climate is concerned, will probably be found in southern and central California, where rains become plentiful especially in the winter half of the year, in the south-eastern quarter of the USA, throughout the Mediterranean region but especially along the north African shores, in parts of India (although the monsoon rains are much weaker than in our own time and much of northern India could well be semi-arid), and in southern China. Lower sea-surface temperature in the tropics means a reduced frequency and intensity of hurricanes in the Caribbean region, and of typhoons and cyclones in the western Pacific. South of the equator, cool temperate climates with adequate rainfall are likely to be found in central Chile, much of South Africa and the southern half of Australia, and in New Zealand's North Island.

This disturbing future world would struggle to support half the year 2000's six billion people at the present-day level of technology. Such a scenario invites human intervention of some kind. Of course, it is easy to forget the timescale we are considering. Human beings have been capable of scientific thought for little more than 5000 years; here we are talking about a period almost 12 times as long, some 58,000 years – more than 2000 generations – in the future. This is hardly the place to predict what advances in scientific thought and technological expertise might have come to pass by then.

Inadvertent anthropogenic climate change in the next 1000 years

Although much less dramatic than the deterioration expected by the next glacial maximum, we have seen that natural climatic changes predicted for the coming millennium are likely to have serious social and economic consequences. And it is during this period that changes triggered by human activity will, in all probability, react with those natural changes.

Some might say that so-called 'global warming' should be encouraged because it will counter the natural cooling predicted during the next several thousand years. This might, just, be an argument worth making if we could regulate the changes in global and regional climate triggered by human activity. Sadly, such technological precision is likely to be beyond our capabilities for a very long time yet. At the moment we are in the position of a chef having just one setting for his oven – full on – and we do not have the luxury of being able to take the dish out of the oven before it is burnt to a cinder. Politicians at international conferences fail to agree to turn the heat down by even one graduation mark on the regulator. The rate of warming during the last two decades is five to ten times more than would be required to counterbalance the natural cooling forecast by Berger's model during the coming 1000 years.

The problem is exacerbated by several characteristics of the global climate machine which are at present poorly understood. These include feedback effects between atmospheric changes, the ocean currents, and the polar ice-caps, and these in turn will create different sorts of climatic change in different parts of the world. Even a regulated system which is calculated to maintain some sort of global climatic equilibrium could still cause catastrophic regional changes.

Because our knowledge of these mechanisms is still so restricted, we could not predict the outcome of human activity on the planet's climate system even if we knew how much carbon dioxide human beings will release into the atmosphere, at what rate, and over what period. But that need not prevent us from making a broad qualitative assessment of the various possibilities which face us. Assuming we do not succeed in developing a satisfactory regulatory system, there are four possible outcomes:

1 Our estimates of anthropogenic climate change are too pessimistic, the present warming phase is short-lived, the global system is able to absorb the effects of higher concentrations of carbon dioxide in the atmosphere, and the pattern of glacial and interglacial periods would continue uninterrupted.
2 A fairly short period of global warming lasting 1000 to 2000 years is superseded by a return to the familiar cycle of glacial and interglacial phases.
3 A long period of global warming lasting for, say, 10,000 years leaves so much residual warmth in the atmosphere that the next glacial period is considerably subdued.
4 The global system is radically altered by a prolonged period of warming leaving the Arctic basin largely unglaciated, which in turn would forestall renewed glaciation even during the most favourable phases of the astronomical cycles; this would produce an irreversible warming, the so-called 'runaway greenhouse effect'.

There is a plethora of scenarios concerning the emission of

greenhouse gases in future decades and centuries – more scenarios than there are experts, in fact. International panels and individual climatologists usually consider several possibilities in their publications, then every two or three years they revise them. Some scenarios are also published by those who pretend to be more expert than they are; yet others are produced with political considerations – sometimes spoken, frequently not – swamping the scientific content. There is, however, a finite limit to the amount of carbon dioxide (just one of several 'greenhouse' gases, it should be remembered) that can be released, and this limit could be reached if we continue to extract fossil fuels (coal, gas and oil) and burn forests until there are none left.

The cynical view is that this is more likely than the other extreme scenario – that we stop using fossil fuels and stop deforestation during the very near future. Nevertheless, it is instructive to examine the repercussions of these two contrasting situations.

Carbon dioxide forms roughly 0.0375 per cent of the Earth's atmosphere; another way of expressing this is 375 ppmv (parts per million by volume). Concentration of carbon dioxide was about 275 ppmv before the industrial revolution, and molecules of the gas have an average residence time in the atmosphere of 200–300 years before being reabsorbed into the biosphere. Bearing in mind that there is still much uncertainty in these predictions, computer models of the planet's carbon cycle suggest that, if we stopped carbon dioxide emissions forthwith, concentrations would continue to increase to approximately 650 ppmv by 2300, and would take something like 5000 years to return to pre-industrial revolution levels. Were we to consume all fuel reserves and forests at an accelerating rate, carbon dioxide could reach something like 2500 ppmv (almost ten times the level in 1800) by the year 2400, and the model predicts that it would take over one million years to return to that eighteenth century baseline.

The Intergovernmental Panel on Climate Change confines its deliberation to the next one hundred years, and thus most of the discussion in the world's news media is restricted to this period. Thus we hear very little about long-term climate-change scenarios. These may be of little direct interest to those of us alive at the moment, but they will be of vital concern to our descendants. The legacy our twentieth- and twenty-first-century civilization is likely to leave is, if our imperfect models are anywhere near correct, not a proud one. All the longer term scenarios indicate that the world's climate fails to return its pre-industrial revolution base even when carbon dioxide levels have declined to that particular threshold. With a peak concentration of 650 ppmv in 2300, it is estimated that global temperature would peak at 5 to 8°C (9 to 13°F) above the eighteenth-century base during the 2400s, dropping back 2 or 3°C by 3000, but remaining about 2°C above the baseline even 10,000 years in the future. In the worst case scenario, with carbon dioxide reaching 2500 ppmv by the year 2400, global atmospheric temperature is predicted to rise some 10 to 15°C (18 to 27°F) above eighteenth-century levels during the 2500s, dropping back 5 or 6°C by 3000, but remaining roughly 5°C above the baseline for many thousands of years into the future.

An increase in global atmospheric temperature of 10°C (let alone 15) would make the world considerably warmer than it was even during the Cretaceous geological period, when both the Arctic and the Antarctic were ice-free and sea-level was 250 metres (over 800 feet) higher than it is today.

We have no way of knowing whether western Europe's climate would be affected to a greater or lesser degree than other parts of the planet. However, we may briefly speculate on Britain's climate in the year 2500 on the assumption that the region experiences a level of warming equal to the global mean – that is, by at least 10°C (18°F). Average daytime temperatures in winter would range from 20°C (68°F) in Cornwall to 15°C (59°F) in north-east Scotland, snow and frost would be largely confined to the Scottish Highlands, and even there would be fleeting and infrequent. In summer, temperatures in southern England would be around 25°C (77°F) at night and 30–35°C (85–95°F) by day, occasionally reaching 45°C (113°F) on the hottest days, and even in northern Scotland 25°C (77°F) would be exceeded on most afternoons.

On the same basis, average daytime maximum temperature in New York would vary between 15°C (59°F) in January and 40°C (104°F) in July, peaking at 50°C (122°F) in summer heatwaves, while Houston would range from 30°C (86°F) in January to 45°C (113°F) on an average July day. Summer temperatures in Sydney, Melbourne, Cape Town and Johannesburg would all average 35–40°C (95–104°F), occasionally reaching 55°C (131°F) on the hottest days.

What effect would global warming have on predictions of the next ice age? It all depends on the degree of global warming. However, André Berger has run the Louvain-la-Neuve model simulating an initial warming sufficient to remove all ice – including the entire Greenland ice-cap – from the northern hemisphere. The Antarctic ice-sheet, however, remains. This time, the anthropogenic warming counters the natural cooling observed in the original model run until 15,000 years into the future. Present day northern hemisphere ice volume is not reached again until AD 20,000 to 25,000, but after that there is a progressive cooling, and the Earth reaches maximum glaciation around 60,000 to 65,000 years from now – slightly delayed compared with the first model run. But the intensity of this future ice age is much reduced, with approximately 30 per cent less ice in the Arctic region, and it is also likely to be of shorter duration.

The detail obtained from this simulation must be highly speculative. However, it is reasonable to suggest that a dramatic period of man-made global warming during the next 500 years would result in a delayed, shortened and weakened glacial phase some 60,000 years in the future.

Appendix

This appendix contains a selection of notable weather and events across the world during the last 1000 years. Some were destructive, some extreme, some merely interesting or unusual. Estimated costs of damage are converted to 2002 prices.

1164 North-west Germany, February. A major North Sea gale resulted in a catastrophic storm surge, flooding large areas of coastal lowland. Known as 'Die Julianenflut', at least 20,000 died.

1219 Denmark, north-west Germany, January. A storm surge flooded large areas of west Jutland and Schleswig-Holstein, with some areas permanently lost to the sea. There was also extensive loss of life.

1238 In England this was arguably the hottest and driest summer of the millennium; the heat and drought were most strongly marked during June and July.

1281 Netherlands, January. A North Sea storm surge piled into the Zuider Zee, flooding large tracts of adjacent land, much of which was below sea-level. An estimated 80,000 people were drowned.

1315 This could well have been the worst summer of the entire millennium in England. Heavy rain and cold winds persisted from June until November. Crops failed to ripen, and serious food shortages followed.

1342 Danube valley, Austria, March. The Great Danube flood; 6000 died.

1362 North-west Germany and the Netherlands, January. The greatest North Sea flood in history, known as

'Die Grosse Manndränke' or 'The Great Drowning'. Coastal flooding affected a broad zone from Schleswig-Holstein to Friesland. Estimates of the numbers drowned range from 100,000 to 500,000. Gales damaged many churches in eastern and southern England, with belfries damaged and spires toppled.

1421 Netherlands, November. A storm crossing the North Sea led to storm surges along the Dutch coast, especially in the Rhine/Scheldt delta and to a lesser extent around the Zuider Zee. This was the worst catastrophe in the Netherlands' long history of flood disasters, with over 100,000 dead.

1430s Six of the seven winters from 1430–31 to 1436–37 were severely cold, with widespread ice and snow across western and central Europe. The 1430s was, alongside the 1690s, the snowiest and frostiest decade of the millennium.

1436 North-west Germany, January. The greatest flood of the fifteenth century with much loss of life. The island of Nordstrand was completely cut in two.

1570 Belgium, Netherlands, Germany, Denmark, November 1–2. Known as the All Saints' Flood, there was loss of life in all these countries. Some contemporary accounts put the total lost at 400,000, but modern estimates suggest fewer than 50,000.

1634 North-west Germany, Denmark, October. Another North Sea storm surge resulted in losses of land almost as large as those of 1362.

1657 England, December 11 to March 21 1658. Snow reputedly lay over much of England for 101 consecutive days.

1683–4 The coldest and snowiest winter in England since the beginning of rudimentary instrumental records in 1659. The *Lorna Doone* winter.

1689 Montafon, Austria, February. An avalanche killed 300.

1694 Culbin, Nairn, Scotland, October. The estate and village of Culbin, between the towns of Nairn and Forres, was completely buried in sand up to a depth of 15 metres (50 feet) followed by a prolonged north-westerly gale which moved coastal sand-dunes inland. Sixteen farms together with the manor house vanished under the sand.

1703 UK, France, Benelux, Denmark, Sweden. November 26–27. Arguably the most violent gale ever felt inland in southern England. Thousands of buildings were damaged, windmills collapsed, and the Eddystone lighthouse was destroyed. Total loss of life was around 8000, over half at sea.

1737 Calcutta, India, October. A cyclone travelling northward through the Bay of Bengal made landfall south of Calcutta. The associated storm surge flooded much of the low-lying land between the city and the coastline, and some 300,000 people were drowned – India's worst cyclone disaster.

1740 UK and Netherlands, January 11. The coldest day in one of the coldest winters of the millennium in northwest Europe. Daytime temperature estimated at minus 20°C (minus 4°F) in Amsterdam and minus 9°C (16°F) in London, accompanied by an easterly gale.

1780 Windward Islands, October. Some 24,000 died in a hurricane which travelled the length of the island chain. Worst hit were St Vincent, St Lucia, Dominica, Guadeloupe, and Martinique.

1786 Japan, summer. Widespread floods. An estimated 30,000 died.

1791 Cuba, June. The most deadly hurricane to hit the country. Over 3000 died.

1814 London, UK, February 1–4. Last Frost Fair to be held on the river Thames in London.

1816 North America and Europe, summer. This was one of the coldest summers ever recorded on both sides of the Atlantic, so much so that 1816 was dubbed 'the year without a summer'. The cold weather was linked with a major volcanic eruption the previous year.

1824 St Petersburg, Russia, November. A ferocious westerly gale piled water from the Baltic Sea into the Gulf of Finland, creating a huge storm surge in what was then the Russian imperial capital. Thousands of buildings were flooded and 569 people were drowned.

1831 Iraq, February. The Great Euphrates Flood; some 15,000 died.

1831 Barbados, September. The island's worst hurricane disaster. Winds flattened thousands of buildings, killing 1500 people – almost two per cent of the population.

1861 Cherrapunji, India, July. The 12 month rainfall between August 1860 and July 1861 at this colonial hill station in the Khasi hills totalled 26462 millimetres (1041.8 inches). It is probably easier to appreciate this quantity if it is expressed as 26.46 metres (nearly 87 feet).

1871 Wisconsin, USA, October 8. Forest fires in the Pestigo area killed 1500.

1875 Garonne valley, France, June. Torrential rains over south-western France, especially in the foothills of the Pyrenees, led to a sudden surge of water entering the Garonne river. The resulting floods killed 400 people.

1876 Bakarganj, Bangladesh, October. A cyclone and storm surge killed an estimated 200,000 to 250,000, the worst human disaster of the 1800s.

1879 Scotland, December 28. The Tay Bridge Disaster occurred on this date, when the newly-opened railway bridge over the Firth of Tay collapsed during the fierce westerly gale. A train was crossing the bridge at the time, and all the passengers were lost.

1881 Southern England, January 18–19. One of the UK's most severe snowstorms on record. On the Isle of Wight level snow was 1.2 metres (four feet) deep, with drifts 4.5 metres (15 feet) high.

1881 China, Haifeng, October. China's worst cyclone disaster. Approximately 300,000 died in this region, some 160 kilometres (100 miles) north-east of Hong Kong.

1882 Bombay, India, June. A cyclone was responsible for the loss of over 100,000 lives, the greatest death toll on the west coast of India for a meteorological event.

1887 California, USA. February 5. San Francisco's greatest ever snowfall with 10 centimetres (4 inches) downtown, and 18 centimetres (7 inches) on the hills.

1889 Pennsylvania, USA, May 31. The Johnstown Flood happened when a heavy rainstorm triggered a dam failure above this small Pennsylvania township. The resultant flash flood wrecked the town and drowned 2100 people. This was the greatest death toll for a flood in the USA.

1891 Southern England, March 9–11. This snowstorm was particularly severe in Cornwall and Devon, where it was accompanied by a severe easterly gale. 220 people died, 6000 sheep perished, and 14 railway trains were stranded in deep snowdrifts in the county of Devon alone.

1895 Braemar, Scotland, Febraury 11. The UK's lowest temperature on record of −27.2°C (−17°F) was first recorded on this date. It was equalled in 1982 and 1995.

1897 Leyte, Philippines, October. The relatively small island of Leyte, just north of Mindanao, was laid waste by the Philippines' worst ever cycone. The death toll exceeded 10,000.

1898 Puerto Rico, August. The San Ciriaco Hurricane, the worst in the island's history, flattened large areas and killed 3369 people.

1900 Galveston, USA, Sept 8. This was the USA's greatest hurricane disaster. The huge storm surge sweeping in from the Gulf of Mexico virtually destroyed the city of Galveston, wrecking over 3000 homes and killing about 8000 people.

1906 Hong Kong, September. Some 10,000 died in the colony's worst cyclone disaster.

1913 Death Valley, California, USA, July 10. The USA's highest temperature on record of 56.7°C (134°F) was logged at Greenland Ranch.

1913 South Wales, UK, October 27. Tornadoes were observed at a number of places in south-west England, Wales and Cheshire, but the one which crossed the small town of Edwardsville, just north of Cardiff, killed six people.

1915 Fort Yukon, Alaska, June 27. A record temperature of 38°C (100°F).

1918 Minnesota and Wisconsin, USA, October. Forest fires killed over 1000.

1921 Thrall, Texas, USA, September 10. The USA's greatest 24-hour rainfall on record was 97 centimetres (38.2 inches).

1922 Washington, DC, USA, January 28. The Knickerbocker Storm – 70 centimetres (28 inches) of snow fell in the city, resulting in the roof of the Knickerbocker movie theater collapsing, killing about 100 people.

1922 Al'Aziziyah, Libya, September 13. The highest recorded temperature in the world of 58°C (136.4°F) on this date.

1925 Missouri/Illinois/Indiana, USA, March 28. The Great Tri-State tornado was the biggest, longest-lasting, and most destructive in the USA's history. The death toll was 695, with 45 further deaths from other tornadoes on the same day.

1927 Danube flood plain, Romania, January. Heavy rain and snow-melt swelled the waters flowing into the Danube, leading to extensive flooding along its course. Over 1000 died.

1927 Mostaganem, Algeria, November. Torrential rain over the Atlas mountains triggered a flash flood which tore through this large town in north-western Algeria, killing over 3000.

1928 Warsaw, Poland, July 6. A rare tornado tore a path through the Polish capital, damaging hundreds of buildings, and killing 82 people.

1930 to 1940 The Great Plains, USA. These were the Dust Bowl Days, with drought prevailing, more or less severely, throughout the decade. Worst hit were Colorado, New Mexico, Oklahoma, Kansas and Texas, and at the peak of the drought in early 1936 the Dust Bowl covered 200,000 square kilometres (80,000 square miles). Hundreds of farmers went out of business as their crops withered.

1931 China, July and August. Monsoon floods, worst along the Yangtze river, were responsible for 1.4 million deaths.

1932 California, USA, January 16. Los Angeles' heaviest snowfall on record – it lay 2.5 centimetres (1 inch) deep at the Civic Center.

1934 Mount Washington, New Hampshire, USA, April 12. A gust of 231 mph (100 m/sec) is accepted as the

USA's highest authenticated wind speed.

1940 Southern England, January 26–29. The worst ice-storm in the UK's history paralysed a large part of the country extending from Hampshire and Sussex in the south to north-east Wales and Shropshire in the north. Elsewhere there was heavy snow.

1947 UK, January 21–March 16. Britain endured a severely cold and very snowy winter while still recovering from the ravages of war. A series of ferocious snowstorms in early March was followed by widespread floods and damaging gales.

1947 Yukon, Canada, February 3. At Snag, Yukon Territory, the temperature fell to −63°C (−81.4°F), the lowest on record for North America.

1949 Guatemala, October. A week of violent rainstorms associated with a stalled tropical cyclone led to flash floods and mudslides throughout eastern Guatemala. The death toll exceeded 40,000.

1950 Canada, USA, western Europe, September 22–28. Forest fires in Alberta led to a huge smoke pall over northern and eastern parts of the USA. A few days later the sun and moon appeared blue in Britain and other parts of western and central Europe.

1951 USA, January 28–February 1. An ice-storm hit a broad swath extending from Texas to West Virginia. Clear ice up to 10 centimetres (4 inches) thick formed on all surfaces in parts of Tennessee, paralysing all outside activity, and cutting power supplies to much of the state. 25 people died, over 500 suffered injuries in accidents on the ice, and damage was estimated at US $200 million.

1952 Devon, UK, August 15–16. The small coastal town of Lynmouth was partly destroyed by a severe flood following an all-night downpour over the adjacent high ground of Exmoor. 34 lives were lost and 420 people were made homeless. At Simonstown, on Exmoor, 231 millimetres (9.1 inches) of rain fell.

1952 London, UK, December 2–6. This was the UK's single most deadly meteorological disaster. The Great London Smog was responsible for some 6000 people dying prematurely. Concentrations of sulphur dioxide and smoke were up to 60 times the normal background level, and dense fog persisted for the best part of four days with horizontal visibility often below seven metres (25 yards).

1953 Netherlands and UK, January 31–February 1. By far the most destructive North Sea storm surge of modern times. 2000 were drowned in the Netherlands and almost 300 along the east coast of England. It triggered a 30-year civil engineering project designed to keep the entire Dutch population safe from future North Sea gales. Insurance losses were about US $6 billion.

1954 Qazvin, Iran, August. A violent thunderstorm over the Elburz mountains triggered a flash flood which swept through this city, to the north-west of Tehran, killing 10,000.

1955 Dorset, UK, July 18. A total of 279.4 millimetres (11 inches) of rain fell in 10 hours at Martinstown, near Dorchester. This is Britain's largest ever fall in one day.

1959 California, USA, February 13–19. The USA's greatest ever snowstorm dumped a total of 480 centimetres (189 inches) of snow at Shasta Ski Bowl.

1959 Northern China, July and August. Widespread floods, chiefly along the Hang-He (Yellow River) during an excessively wet summer monsoon. An estimated two million deaths was the highest number ever for a meteorological disaster in the world.

1959 Ise Bay, Japan, September. Typhoon Vera swept across this region, south of Nagoya, claiming 5098 lives, Japan's greatest death toll in a typhoon in modern times.

1962 Barcelona, Spain, September 27. Spain's worst natural disaster of the twentieth century, with 474 people drowned as flash floods tore through the country's second largest city.

1962–1963 UK, December 22–March 4. The most severe winter in Britain for over 200 years, with snow covering the ground for 69 consecutive days beginning December 26 in many places. The temperature failed to exceed 3°C (37°F) at several sites between December 22 and March 1.

1963 Longarone, north-eastern Italy. A dam failed following the very wet summer, and a wall of water crashed through this small town. Most of the population succumbed, the death toll reaching 1896.

1963 Haiti, October. Hurricane Flora hit the island, killing 5100 people.

1964 Oymyakon, Siberia, February. The lowest temperature ever recorded in the northern hemisphere was minus 71.1°C (minus 96°F).

1964 Karachi, Pakistan, December. Pakistan's worst cyclone disaster with 10,000 deaths.

1969 Central Tunisia, September. Severe flooding led to 542 deaths.

1970 Hungary and Romania, June. Prolonged heavy rains during several weeks, especially over the Carpathian mountains, fed into the river Tisza which broke its banks and flooded neighbouring farmland in eastern Hungary. Some 300 people died in Hungary and a further 200 in northern Romania.

1970 Kansas, USA, September 3. A hailstone which fell at Coffeyville, Kansas, measured 44 centimetres (17.5

inches) in circumference and weighed 758 grams (1.67 pounds). This is the largest authenticated hailstone known to have fallen in the USA.

1970 Bangladesh, November 13. A cyclone hit the eastern delta region, and the associated storm surge, estimated at 4 to 5 metres (13 to 16 feet), swamped over 250 square kilometres (100 square miles) of land. The death toll was variously put at anything between 300,000 and 750,000 – the greatest casualty list for any single meteorological event in the twentieth century.

1973 South-east Spain, October. Widespread floods in the cities of Granada, Murcia and Almeria. The death toll was 350, and losses were put at US $300 million.

1974 Darwin, Australia, December 25. Cyclone Tracy struck during the early hours, killing 65 people and costing US $2 billion. Extensive damage revealed that most buildings in the town were built to insufficiently high standards for an area at risk from cyclones.

1975 UK, Benelux, Germany, the Baltic, January 2–3. A vigorous Atlantic depression nicknamed 'Capella' screamed across northern Britain, Denmark, and into the Baltic, with gusts as high as 115 mph (50 m/sec), which caused extensive damage to property, disrupted power supply, and dislocated transport. 82 people died, and insurance losses were put at US $2.6 billion.

1975 London, UK, August 14. A tremendous thunder and hail storm broke across the fashionable north London suburbs of Hampstead and Highgate, and floodwaters paralysed the Underground railway system. At Hampstead observatory 171 millimetres (6.72 inches) of rain fell in 150 minutes.

1976 UK, Summer. The hottest and driest summer on record was the culmination of 16 months of drought

– the worst since rainfall records began in England in the seventeenth century. Water for industrial and domestic purposes was rationed, a minister for drought was appointed, and heath fires burned substantial areas of southern England. The temperature exceeded 32°C (90°F) on 15 consecutive days in late June and early July.

1977 Florida, USA, January 19. Light snow showers fell throughout the state, including Miami. Snowflakes were even observed at the extreme southern tip of the mainland.

1977 Eastern USA, Jan/Feb. Prolonged severe winter. Insurance losses estimated at US $5.6 billion.

1977 Madaripur, Bangladesh, April. A swarm of tornadoes cut a swath through this region to the south-west of the capital, Dakha. Over 900 died, the highest number of tornado-related deaths for a single day anywhere in the world.

1978 Baden-Württemberg, Germany. The worst floods in modern times in Germany took 300 lives.

1979 British Isles, August 13–14. A vigorous Atlantic depression swept across the British Isles with gusts as high as 85 mph. This severe gale coincided with the Fastnet Race, the culmination of the Admiral's Cup competition. Of the 303 starters, only 85 finished, 194 retired, and 23 were abandoned, of which five sank. A total of 136 crew had to be rescued, but 15 drowned.

1981 Denmark, November 24. Intense depression crossed the country. Winds gusting over 100 mph (45 m/sec) caused widespread damage. Nine people died. A rig at the Argyll oil-field in the central North Sea lost its anchors and drifted out of control for 48 hours; waves were over 15 metres (50 feet) high.

1982 Braemar, Scotland, January 10. The UK's lowest temperature of –27.2°C (–17°F) was equalled on this date. The previous day's maximum reading at Braemar was –19°C (–2°F).

1982 Peru and Ecuador, January to April. Unprecedented rains associated with the most intense El Niño on record led to widespread flooding, which claimed 500 lives and caused US $1.4 billion worth of damage.

1983 Victoria, Australia, February. Widespread brush fires; 75 people died, with US $460 worth of million damage.

1983 Vostok, Antarctica, July 21. The officially-recognized lowest temperature anywhere in the world was –89.2°C (–138.6°F).

1984 Munich, Germany, July 12. The most expensive day's weather in Germany's history occurred when a hailstorm struck the country's third largest city. No-one was killed, but insurance losses exceeded US $2 billion.

1985 Cairn Gorm, Scotland, March 20. The UK's highest ever wind speed, 171mph (74 m/sec) was recorded on the summit of Cairn Gorm in the Grampian mountains.

1987 Southern England, January 12. Maximum temperature on this day was –9°C (16°F) at several sites in Devon, Surrey, Kent and Sussex, accompanied by a brisk north-easterly wind. These are the lowest daytime temperatures ever recorded in southern England. Some 50 centimetres (20 inches) of snow fell in west Kent and north-east Surrey.

1987 Natal, South Africa, September. Flash floods killed 487, and insurance losses were estimated at US $1.5 billion.

1987 UK, October 15–16. 18 people died in the UK's most famous windstorm of the twentieth century. Only the south-eastern quarter of the country was affected, but this included London where all commercial activity ceased for 24 hours. Famously unpredicted by weather forecasters, gusts of 90–110 mph (40–50 m/sec) were unprecedented in the region, and millions of trees were uprooted. Northern France was also badly hit. Insurance losses were US $3.4 billion.

1988 Jamaica and Mexico, September 12–17. Hurricane Gilbert was the most intense tropical storm ever recorded in the western hemisphere. Most of Jamaica and the north coast of the Yucatan were laid waste as sustained winds reached 175 mph (75 m/sec) with peak gusts of 205 mph (90 m/sec).

1990 UK, January 25. The so-called 'Burns' Day Storm' hit all parts of England and Wales with peak gusts of 108 mph (47 m/sec) causing chaos on road and rail, and widespread structural damage. The death toll was 47.

1990 Cheltenham, UK, August 3. The highest temperature ever officially recorded in the UK was 37.1°C (98.8°F).

1991 Florida, USA, August. Hurricane Andrew swept across Dade County in the south of the state, causing much damage in this heavily populated district. This was the USA's most costly meteorological disaster, with insurance losses in excess of US $4 billion. Mercifully, the casualty list was relatively short.

1995 Chicago, USA, July/August. Prolonged heatwave accompanied by high humidity. Over 700 heat-related deaths.

1995 Altnaharra, Scotland, December 30. The lowest temperature on record in the UK of −27.2°C (−17°F) was logged for the third time on this date.

1998 Guatemala, Honduras and Nicaragua, October. Hurricane Mitch caused considerable damage when it made landfall courtesy of winds in excess of 150 mph (70 m/sec) and a huge tidal surge. But the catastrophe which followed was triggered by a prolonged downpour of rain which led to extensive floods and mudslides. Over 12,000 died.

1999 Orissa, India, October 31. Probably the most intense cyclone ever recorded crossing the Indian coastline, with sustained winds of 160 mph (70 m/sec). Despite reasonably good warnings, this remote coastal region was only partially evacuated and over 10,000 people died.

1999 France, December 26–28. Two intense depressions crossed the country with gusts of 100–120 mph (45–55 m/sec), causing unprecedented damage and almost 100 deaths. It was described as France's worst peace-time disaster. The first storm hit northern France on December 26, the second crossed central France overnight 27–28.

2001 Buffalo, USA, December 24–30. Buffalo is recognised as one of the world's snowiest cities, and a series of heavy storms led to accumulations of around two metres (84 inches) by the end of the year.

Bibliography

A selection of mainly recent publications which will take the interested reader further into those aspects of weather and climate discussed in this book. Some of these provide a scientific background which the lay reader may find difficult.

Barry, R.G. and Chorley, R.J. *Atmosphere, Weather and Climate*, London and New York, Routledge, 1995.

Bate, R. (ed.) *Global Warming, The Continuing Debate*, London, The European Science and Environment Forum, 1998.

Bradley, R.S. and Jones, P.D. *Climate since AD 1500*, London and New York, Routledge, 1992.

Burroughs, W.J. *Climate Change: A Multidisciplinary Approach*, Cambridge, Cambridge University Press, 2001.

Burroughs, W.J. and Lynagh, N. *Maritime Weather and Climate*, London, Wetherby, 1999.

C.S.I.R.O. (eds.) *The Australian Environment*, Melbourne, Melbourne University Press, 1960.

Chandler, T.J. and Gregory, S. (eds.) *The Climate of the British Isles*, London and New York, Longman, 1976.

Changnon, S.A. *El Niño 1997–1998: The Climate Event of the Century*, New York, Oxford University Press, 2000.

Eden, P. *Weatherwise*, London, Macmillan, 1995.

Ellig, J. *Set Fair: A Gradualist Proposal for Privatising Weather Forecasting*, London, The Social Affairs Unit, 1988.

Emsley, J. (ed.) *The Global Warming Debate*, London, The European Science and Environment Forum, 1996.

Fleming, J.R. *Historical Perspectives on Climate Change*, New York, Oxford University Press, 1998.

Garnier, B.J. *The Climate of New Zealand*, London, Edward Arnold, 1958.

Gerholm, T.R. (ed) *Climate Policy after Kyoto*, Brentwood, UK, Multi-Science Publishing Company, 1999.

Giambelluca, T.W. and Henderson-Sellers, A. *Climate Change: Developing Southern Hemisphere Perspectives*, Chichester, UK, John Wiley, 1996.

Glantz, M.H. *Currents of Change: Impacts of El Niño and La Niña*, Cambridge, Cambridge University Press, 2001.

Grove, J.M. *The Little Ice Age*, London and New York, Methuen, 1988.

Harrington, C.R. (ed) *The Year Without a Summer? World Climate in 1816*, Ottawa, Canadian Museum of Nature, 1992.

Harvey, L.D.D. *Climate and Global Environmental Change*, Harlow, UK, Prentice Hall, 1999.

Houghton, J.T. *et al.* (eds) *Climate Change 2001, the Scientific Basis.* Contribution to the Third Assessment Report of the Intergovernmental Panel on Climate Change, Cambridge, Cambridge University Press, 2001.

Huber, B.T. *et al.* (eds) *Warm Climates in Earth History*, Cambridge, Cambridge University Press, 2000.

Hulme, M. and Barrow, E. (eds) *Climates of the British Isles, Present, Past and Future*, London and New York, Routledge, 1997.

Jones, P.D. *et al.* (eds) *History and Climate: Memories of the Future*? Dordrecht, Netherlands, Kluwer Academic Publishers, 2001.

Knowles Middleton, W.E. *Invention of the Meteorological Instruments*, Baltimore, Johns Hopkins University Press, 1969.

Lamb, H.H. *Climate, History, and the Modern World*, London and New York, Routledge, 1995.

Lamb, H.H. *Climate: Present, Past and Future*, vol. 1: *Fundamentals and Climate Now*, London, Methuen, 1972.

Lamb, H.H. *Climate: Present, Past and Future*, vol. 2: *Climate History and the Future*, London, Methuen, 1977.

Lamb, H.H. *Historic Storms of the North Sea, British Isles and Northwest Europe*, Cambridge, Cambridge University Press, 1991.

Le Roy Ladurie, E. *Times of Feast, Times of Famine: A History of Climate since the Year 1000*, New York, Doubleday, 1971.

Leggett, J. *The Carbon War*, London, Penguin, 2000.

Ludlum, D.M. *The American Weather Book*, Boston, Houghton Mifflin, 1982.

Ludlum, D.M. *The Weather Factor*, Boston, American Meteorological Society, 1989.

Manley, G. *Climate and the British Scene*, London, Collins, 1952.

McGregor, G.R. (*et al.*) *Tropical Climatology: an Introduction to the Climates of Low Latitudes*, Chichester, UK, John Wiley, 1998.

Met. Office, The. *Annual Report and Accounts 1999/2000*, London, The Stationery Office, 2000.

NOAA. *National Implementation Plan for Modernization of the National Weather Service for Fiscal Year 1997*, Washington, Department of Commerce, 1996.

Pettersen, S, *Weathering the Storm: Sverre Pettersen, the D-Day Forecast, and the Rise of Modern Meteorology*, Boston, American Meteorological Society, 2001.

Redfern, R. *Origins: The Evolution of Continents, Oceans and Life*, London, Cassell, 2000.

Robinson, P.J. and Henderson-Sellers, A. *Contemporary Climatology*, Harlow, UK, Pearson, 1999.

Rohan, P.K. *The Climate of Ireland*, Dublin, The Stationery Office, 1986.

Ruffner, J.A. and Bair, F.E. *The Weather Almanac*, New York, Avon, 1977.

Smith, K. *Environmental Hazards: Assessing Risk and Reducing Disaster*, London and New York, Routledge, 2000.

Strangeways, I. *Measuring the Natural Environment*, Cambridge, Cambridge University Press, 2000.

Thomas, M.K. *Climatological Atlas of Canada*, Ottawa, National Research Council, 1953.

Tyson, P.D. and Preston-Whyte, R.A. *The Weather and Climate of Southern Africa*, Cape Town, Oxford University Press Southern Africa, 2000.

US Global Change Research Program. *Climate Change Impacts on the United States*, Cambridge, Cambridge University Press, 2001.

Williams, M.A.J. *et al. Quaternary Environments*, London, Edward Arnold, 1993.

Yoshino, M. *et al.* (eds) *Climates and Societies – A Climatological Perspective*, Dordrecht, Netherlands, Kluwer Academic Publishers, 1997.

Index